Essener Geographische Arbeiten 33

Anja Scheffers

Paleotsunamis in the Caribbean

Field evidences and datings from Aruba, Curaçao and Bonaire

33 ESSENER GEOGRAPHISCHE ARBEITEN

Herausgeber:

Gerhard Henkel, Dieter Kelletat, Frank Schäbitz, Wolfgang Trautmann und Hans-Werner Wehling
Institut für Geographie, Universität Essen

Schriftleitung: Gudrun Reichert

Die Deutsche Bibliothek - CIP-Einheitsaufnahme

Anja Scheffers:

Paleotsunamis in the Caribbean

Field Evidences and Datings from Aruba, Curaçao and Bonaire / Anja Scheffers. - 1. Auflage - Essen: Selbstverlag, 2002 (Essener Geographische Arbeiten, Band 33)
ISBN 3-9803484-9-0

1. Auflage August 2002
© Institut für Geographie, Universität Essen 2002
Druck: SLC-GmbH, Essen
Alle Rechte vorbehalten

ISBN 3-9803484-9-0

ESSENER GEOGRAPHISCHE ARBEITEN 33

Für Emile

ESSENER GEOGRAPHISCHE ARBEITEN   Band 33   Essen 2002

Paleotsunamis in the Caribbean
Field Evidences and Datings from Aruba, Curaçao and Bonaire

ANJA SCHEFFERS*

CONTENTS

Contents 1

Abstract 3

Kurzfassung 4

Acknowledgments 5

1. Introduction – Aim of the Study 5

2. Regional Setting and Physical Environment 7
   2.1 Regional Setting 7
   2.2 History 11
   2.3 Climate 11
   2.4 Geodynamics and Bathymetry on a Regional and Caribbean-wide Scale 14
   2.5 Geology and Geomorphology 17
      2.5.1 Coastal Geomorphology 23
      2.5.2 Recent Distribution of Coral Reefs 30
   2.6 Soils and Vegetation 30
      2.6.1 Soils 30
      2.6.2 Vegetation 31
   2.7 Landuse and Infrastructure 33
      2.7.1 Landuse 33
      2.7.2 Economy and Infrastructure 33

3. Hurricanes and Tsunamis: General Aspects and their Relation to
Coastal Forming Processes 36
   3.1 Tsunamis – General Aspects 36
      3.1.1 Tsunami Definition 36
      3.1.2 Tsunami Magnitude Scales 36
      3.1.3 Tsunami Characteristics 38
      3.1.4 Tsunami Generation 41
      3.1.5 Spatial Distribution of Tsunami Impacts on Different Temporal Scales 41
   3.2 Hurricanes – General Aspects 42
      3.2.1 Hurricane Definition 42
      3.2.2 Hurricane Intensity Scale 45
      3.2.3 Hurricane Characteristics 45
   3.3 Hurricanes or Tsunamis – Potential Capacity and Field Observations of Debris Deposition 51
      3.3.1 Hydrodynamic Aspects of Boulder Movement 51
      3.3.2 Hurricanes and Tsunamis – Worldwide 57
      3.3.3 Hurricanes and Tsunamis in the Intra Americas Sea (IAS) 59
      3.3.4 Hurricanes and Tsunamis on Curaçao, Bonaire and Aruba 66

---

* Anja Scheffers, University of Essen, FB 9 - Institut of Geographie, Universitätsstr. 15, D - 45117 Essen / e-mail: anja.scheffers@uni-essen.de

4. Materials and Methods	79
4.1 Field Survey	69
4.2 Characterization System	69
4.3 Situmetry and Shape Measurements	69
4.4 Interpretation of Aerial Photography	70
4.5 Relative and Absolute Age Dating of the Deposits	71
4.5.1 Relative Age Dating	71
4.5.2 Absolute Age Dating	71
4.6 Microfossils	72
5. Field Evidences	73
5.1 Aruba	76
5.1.1 The Leeward Coast (S to N)	77
5.1.2 The Windward Coast (N to S)	78
5.1.3 Summary and Conclusion	90
5.2 Curaçao	91
5.2.1 The Leeward Coast (S to N)	93
5.2.1.1 The Probable Genesis of Caracasbaai and the Caracasbaai Slide	103
5.2.2 The Windward Coast (N to S)	105
5.2.3 Summary and Conclusion	117
5.2.4 Klein-Curaçao	118
5.3 Bonaire	119
5.3.1 The Leeward Coast (S to N)	121
5.3.2 Klein-Bonaire	129
5.3.3 The Windward Coast (N to S)	129
5.3.4 Summary and Conclusion	141
5.4 Comparison of the Paleotsunami Deposits of Aruba, Curaçao and Bonaire	143
6. Dating and the Depositional History of the Paleotsunami Events	147
6.1 Relative Age Dating	147
6.2 Absolute Age Dating	148
6.2.1 Biological Differences of the Age Distribution – the Species	153
6.2.2 Geomorphological Differences of the Age Distribution – the Debris Formations	155
6.2.3 Geographical Differences of the Age Distribution – Comparison of the Islands	157
6.2.4 Comparison of Relative and Absolute Ages	162
6.2.5 Future Applications	162
7. Linking of the Paleotsunami Deposits to Source Mechanisms and Runup Heights	163
8. Conclusions	165
List of Figures	167
List of Tables	173
References	174
Publications – Institut of Geography	183

Abstract

Aruba, Curaçao and Bonaire, located north of the Venezuelan coast in the Caribbean Sea, exhibit several attributes that have permitted a detailed regional characterization of the morphology of tsunami deposits. Hitherto, tsunami impacts were unknown for the ABC-islands and the debris formations have been exclusively attributed to hurricane-generated waves.

The accumulations exhibit three main geomorphologic distinct types of paleotsunami debris formations, which have been distinguished as boulder assemblages, rampart formations and ridge formations. Predominantly, the debris deposits have been accumulated on the northeastern sides of the islands, reaching from sealevel to a height of +12 m asl and extending up to 400 m inland. On a regional scale, the extent and amount of tsunami debris weakens from east to west with the highest energy impact on Bonaire in the east and a considerable lower impact on Aruba, the most westerly island. Each formation exhibits a distinct morphology and geographic distribution related to a certain coastal configuration.

Boulder assemblages contain blocks of >100 m^3 in volume and with a weight of up to 281 tons. They occur on all islands with the most impressive evidences on Bonaire and Curaçao, but in general, they are coinciding remarkably often with coastal sections, where the cliff front is nearly perpendicular and the supratidal zone is rather narrow.

If the coastal physiography leads to the development of a rather broad supratidal with a more convex cliff profile, this coastal environment favors the development of rampart formations. They occur likewise on all islands with the most developed ones in northeastern Curaçao and along the east-exposed coastal stretch on Bonaire.

The ridge deposits often follow subsequently to the coastline and surf zone. These ridges occur along the southern, southeastern and western leeward coastlines, where they may extend over several hundred meters with width from 10-50 m and relative heights from 1-3 m.

One key problem concerns the differentiation between a storm-induced or tsunami-induced sedimentary record. During the time period 1605-2000 in total 14 hurricanes and 19 tropical storms of minor intensity passed the islands within the 100 nm zone.

The most significant event in the past was Hurricane *Lenny* in November 1999, an extremely rare hurricane with wind speeds >160 km/h, formed south of Jamaica and moved eastward toward the Lesser Antilles. As a result of the rather unusual track, the islands of Aruba, Bonaire and Curaçao experienced heavy surf conditions along their southwestern coastlines. It can be clearly observed that the magnitude of the paleotsunami events exceeded the impact of hurricane *Lenny* significantly on all three islands.

Relative age indications allow a good estimation of the time range for the minimum and maximum age of the deposits. Overall, we can limit the maximum age range to the Younger Holocene as evident in particular by chemical and biological weathering processes and the spatial relation of the debris formations to the recent sealevel highstand. Beside geomorphic imprints of tsunami occurrence the historical record has to be considered. On the ABC-islands no written or oral sources describing a tsunami impact exist, pinpointing also to a time span of minimum 350-400 years without the occurrence of any severe tsunami event, presumably since the Dutch occupation in 1634 AD or even the occupation by the Spaniards in 1527 AD.

However, the resolution of relative age dating is insufficient to establish a more detailed chronology of the tsunami impacts, so that radiocarbon age determinations from 43 samples were performed from different geomorphologic units (boulders, ramparts, ridges) and on different material (vermetids, coral, gastropods).

These conventional radiocarbon datings supplied a non-calibrated age range from 370 ± 32 to 4222 ± 49 years BP. The uncalibrated age data show a clustering in three main time units around 500 BP, 1500 BP and 3500 BP with intermediate periods of only infrequent or no age values.

The generating mechanisms of paleotsunamis of the described magnitude is unknown, but most likely they are related to seismic activity in the northeastern part of the Antillean Island Arc along the faults of the Caribbean Plate boundaries.

Kurzfassung

Ungewöhnlich verbreitete grobblockige Ablagerungen auf den Inseln Curaçao und Bonaire der Niederländischen Antillen sowie Aruba erlauben, eine differenzierte morphologische Charakterisierung von Tsunamiablagerungen im regionalen Maßstab vorzunehmen. Bislang wurden die Grobschuttablagerungen in der regionalen Literatur ohne detaillierte Bearbeitung der alleinigen Einwirkung von Hurrikanen zugeschrieben und die Möglichkeit einer tsunamigenen Entstehung völlig außer acht gelassen.

Die Ablagerungen finden sich überwiegend auf den nordöstlichen Inselseiten und reichen vom gegenwärtigen Meeresspiegelstand bis in Höhenlagen von +12 m mit einer landwärtigen Reichweite von bis zu 400 m. Im regionalen Maßstab vermindert sich der Umfang und die Dimension der abgelagerten Schutteinheiten von Ost nach West mit der gewaltigsten Einwirkung auf der östlichen Insel Bonaire und deutlich geringeren Auswirkungen auf dem mehr westlich gelegenen Aruba.

Als ein Ergebnis der Feldanalysen werden drei Grobmaterialablagerungen unterschieden: Die weiteste Verbreitung weisen die "ramparts" auf, welche als dekameterbreite und bis zu einem Meter mächtige deutlich durch einen sedimentfreien Streifen vom Kliff abgesetzte Schuttdecken ausgebildet sind. Die Korngrösse erreicht mehrere 100 kg, das Material stammt überwiegend aus dem Supralitoral des jungpleistozänen gehobenen Korallenriffs.

Als "ridges" werden solche Ablagerungen bezeichnet, die eine deutliche Wallform aufweisen. Sie können bis zu +3 m mächtig, über 50 m breit und mehrere Kilometer lang werden. Unmittelbar am Brandungssaum gelegen enthalten auch sie sehr groben Schutt, der aufgrund besserer Zurundung, bioerosiver Überformung und Beimengung gut erhaltener Korallenfragmente eine Herkunft aus dem foreshore-Bereich belegt.

Die dritte Einheit sind Einzelblöcke oder "boulder assemblages" von mehreren bis zu 300 Tonnen Gewicht. Sie sind von eckiger Form, losgerissen aus dem Kliff oder der bench und liegen verstreut innerhalb bzw. gehäuft auch vor den ramparts auf der unteren Riffterrasse in Höhen von +2 bis +12 m.

Herauszustellen ist, dass sich die ABC-Inseln am südlichen Rande der Hurrikanzugbahnen befinden und seit Beginn verlässlicher meteorologischer Beobachtungen (1886) und der 400-jährigen Aufzeichnungen der Seefahrt lediglich alle 100 Jahre ein Hurrikan der Kategorie 2 (von insgesamt 5 der SAFFIR-SIMPSON-Skala) innerhalb von 100 Seemeilen (180 km) passiert. Der Differenzierung der Wirkungsweise von Hurrikanen und Tsunamis im Gelände kam das Auftreten von Hurrikan *Lenny* im November 1999 zugute, der als einziger im Beobachtungszeitraum der letzten 120 Jahre (mit über 900 schweren karibischen Stürmen), wahrscheinlich sogar der letzten 500 Jahre in der bekannten Hurrikangeschichte der Karibik eine West-Ost-Zugbahn aufwies, wodurch die ABC-Inseln durch außergewöhnliche Wellenereignisse betroffen wurden. Dabei belegen die Feldbefunde eindeutig, daß die Größenordnung der Tsunamiablagerungen diejenigen von *Lenny* um ein Vielfaches übersteigen.

Die geomorphologischen und sedimentologischen Feldbefunde, und hier insbesondere chemische und biologische Verwitterungsprozesse sowie die Lagebeziehung der Ablagerungen zum gegenwärtigen Meeresspiegelstand, ergaben eine ganze Reihe von Kriterien für eine relative Alterseinschätzung der Tsunamis von mindestens einigen Jahrhunderten bis zu maximal wenigen Jahrtausenden. Dieser Altereinschätzung entspricht auch die Beziehung der Tsunami-Ablagerungen zu anthropogenen Spuren. Trotz der ständigen Anwesenheit der Kolonialmächte spätestens seit 1634 und einem regen Schiffsverkehr bereits sei dem 16. Jahrhundert gibt es keinerlei Berichte, die mit den großen Tsunamis in Zusammenhang gebracht werden könnten.

Zur absoluten Altersbestimmung wurden 43 karbonatische Proben (Korallen, Vermetiden und Gastropoden) aus den unterschiedlichen geomorphologischen Einheiten der ^{14}C-Methode zugeführt. Die Altersverteilung der nicht kalibrierten Daten reicht von 370 ± 32 bis 4222 ± 49 Jahren BP, wobei klar drei Schwerpunkte um 400-500 BP, ein zweiter um 1500 BP und ein dritter um 3500 BP erkennbar sind, zwischen denen lange Zeitabschnitte mit nur wenigen oder gar keinen Daten liegen.

Noch ist offen, welche Vorgänge die Tsunamis ausgelöst haben, doch dürften diese wahrscheinlich in Zusammenhang stehen mit den seismisch aktiven Regionen entlang der Grenzen der karibischen Platte im nordöstlichen Sektor des Antillenbogens.

Acknowledgments

This research was completed as a doctoral dissertation under the direction of Dieter Kelletat. With his indefatigable support, he made a significant contribution to this work. I would like to give particular thanks to his wife B. Kelletat for her warmhearted personnel input.

Financial support for this research was provided by the German Scientific Community "Deutsche Forschungsgemeinschaft (DFG)", project "Littoral coarse debris on the Netherlands Antilles in the Holocene: Hurricanes or Tsunamis?" (Ke 190/17-1).

The successful completion of the research required the skills and collaboration of many people, with special thanks to the following:

"The Kelletat-Team" with Anne Hager and Gudrun Reichert for conducting the graphic illustrations, technical procedures and literature research;

Leendert Pors (CARMABI) for supporting me with a computer workplace on Curaçao and immense scientific and personal assistance;

Gerhard Schellmann for sending me to Curaçao and that he watched my steps there with a well-disposed smile on his face;

B. Kromer for conducting ^{14}C analyses at the Institut für Umweltphysik, Heidelberger Akademie der Wissenschaften; Paula Reimer, Queen's University of Belfast, for her advice in calibrating questions of the radiocarbon dataset;

J. Lobenstein for completing the foraminifer identifications;

The Island Government of Curaçao and L. Hooi and G. Principal (DROW) for providing the aerial pictures of Curaçao. The Island Government of Bonaire through Peter Montanus (DROB) for providing the aerial pictures of Bonaire.

U. Radtke and his wife D. Dorra for providing help during the initiation of the project and the start of my stay in Curaçao.

To the entire staff of the Carmabi Foundation a great compliment and particular thanks for all the support during minor and bigger difficulties in everyday research life and with bureaucratic obstacles.

I also owe much for the hospitality and assistance provided by the people of Curaçao, Bonaire and Aruba during the field work. *Danki*!

My deepest thanks for the help and encouragement to: Sander Scheffers, my families in Germany and The Netherlands, Stephan Theune and family, and Bärbel Kaiser and family.

1. Introduction – Aim of the Study

Relating a geological deposit to a paleotsunami is a delicate exercise. Worldwide it is debated whether it is possible to differentiate between paleotsunamis and paleocyclones in the record of old sediments. CHAGUÉ-GOFF & GOFF (1999) stated *"while it is easy enough to identify the difference between a large, well documented tsunami and a small, well documented cyclone, there is a significant gray area in the middle"*. A field study related to age determination of the fossil reef terraces (RADTKE et al., 2002) and the sight of the existing geomorphologic literature suggested that it might be possible to study exactly this "gray zone" in the Leeward Netherlands Antilles.

With this in our mind, the DFG-Project "Littoral coarse debris on the Netherlands Antilles in the Holocene: Hurricanes or Tsunamis?" was initiated[1]. During the investigations it became evident that most of the coastal debris deposits of the islands are the result of tsunami impacts. This is important, because hitherto no tsunami records were documented for the ABC-islands, although for the past 500 years, the Caribbean has experienced teletsunamis, tectonic tsunamis, landslide tsunamis and volcanic tsunamis (LANDER & WHITESIDE, 1997). LANDER et al. (1998) compiled a history of 88 reported tsunamis since 1530 for the entire Caribbean region showing similar fatality levels in the last 150 years as in Hawaii, Alaska and the West Coast of the American States combined.

Yet despite concentrated research, many aspects of the nature and the occurrence of paleotsunamis remain incomplete. The event, be it catastrophic or everyday, has passed and all that remains is the memory preserved in the landscape, what is in most cases just an erosional or depositional record. Nevertheless, references to historical events are necessary to understand the nature of past tsunamis in order to learn how they will affect a certain area in the future.

---

[1] *DFG: Deutsche Forschungsgemeinschaft – the most important research organization in Germany.*

It is to be regretted that in most of the literature of paleotsunamis the attention is focussed on the deposition of anomalous sand or mud units and boulder formations are mostly simply ascribed to hurricane or storm effects. This is also the case within the literature subjected to geologic and geomorphologic features of the Leeward Netherlands Antilles. DE BUISONJÉ (1974) describes the deposits as "hurricane rubble" or "coral shingle" and the depositional process as that *"the ridge consists of loose blocks that during extremely stormy weather broke off from the cliff and were accumulated in a distance of some tens of meters from the coastline proper"*.

Curaçao, Bonaire and Aruba exhibit several unique attributes that have permitted a detailed characterization of the morphology and structure of tsunami deposits and the related forming processes.

- The youngest geologic history of the islands is a tectonically quiescent period. During the Quaternary, the islands have undergone a relatively slow vertical uplift, which allows linking the height and distance of the deposits directly to the sealevel, and no neotectonic dislocations have to be considered.

- The islands show a relatively similar stratigraphy and lithology. This facilitates the interpretation of differences or similarities in congruent height-levels. The wide occurrence of carbonate rocks is responsible for a variety of specific geomorphologic features like notches, benches or algal rims, which can be used for relative and absolute dating.

- The study of these islands allow conclusions for the Southern Caribbean over a geographical distance of more than 200 km.

- Due to their geographical position at the southern fringe of the hurricane belt, major tropical storms or hurricanes only occasionally touch the islands. This results in an excellent preservation of coastal deposits and eventually allows analyzing single events, either from storms or tsunamis.

- The limited hurricane impact causes an increased stability of biogenous fine structures of the coastal area with respect to conclusions concerning the relative age of the forms and the intensity of the forming processes.

- Conversely, due to their geographical position the ABC-islands represent a focus-point for tsunami waves generated within the West Indies island arc.

Finally, the need for a better understanding of tsunami hazards in the wider Caribbean region is illustrated by the following statements of the IOCARIBE Tsunami Steering Group of Experts: *"First, the Caribbean is at risk for tsunamis and is essentially unprepared for any such event. Second, since the last major event of the 19th century (St. Croix, 1867) the population of the Caribbean region has increased approximately ten-fold from 3 million to 30 million, and at least half or more people live in the coastal area. Third, Caribbean nations are, for the most part, uninformed of tsunami risks and uninvolved in their warnings. Fourth, government leaders at the highest levels must be informed and be held accountable for the information. And finally, the Epiphany: There's always money for workshops; there's never money for work ! must undergo a paradigm shift"* (MAUL, 1999).

With these considerations in mind, the Leeward Netherlands Antilles have been chosen as a natural laboratory to study the distribution and nature of paleotsunami debris deposits. In contrast to previous studies, which are mainly focussing on historical earthquake records as possible tsunami generating events, this research chooses an inductive conception with studying the spatial distribution and geomorphic evidences related to coarse littoral sediment deposition of Holocene age in the Southern Caribbean. We hope with this research to highlight the necessity of intensified research of paleotsunamis in general and tsunami hazards in the Caribbean. And, in a wider context, to contribute to the establishment of a Caribbean Tsunami Warning System, following the example of the Pacific Tsunami Warning System, which has prevented unnecessary severe damage and loss of life.

The research presented in this study is organized in seven chapters, each of which play an integral part in establishing the depositional history of the tsunami and hurricane debris deposits on the Leeward Netherlands Antilles. Chapter 1 introduces in the overall problematic. The regional setting, the physical environment and the anthropogenic background as far as they are related to coastal questions/problems, are described in Chapter 2. Chapter 3 summarizes the present day knowledge concerning the nature of tsunamis and hurricanes, their energetic potential and the related depositional processes worldwide, within the Caribbean region and for the ABC-islands themselves. The research methods are presented in Chapter 4. Chapter 5 establishes the sedimentology and depositional environments of the tsunami and hurricane debris deposits. Chapter 6 utilizes relative and absolute age dating results to determine the depositional history of the coastal debris deposits. The source mechanisms are briefly discussed in Chapter 7. And Chapter 8 synthesizes the conclusions presented in the previous chapters.

2. Regional Setting and Physical Environment

2.1 Regional Setting

The Leeward Netherlands Antilles consist of five islands: Curaçao and its small satellite Klein-Curaçao, Bonaire with a similar neighbor Klein-Bonaire, and Aruba. The islands are located north of the Venezuelan coast in the Caribbean Sea between 12° and 13° northern latitude and between 68° and 70°04' western longitude. They are lying roughly parallel to the coast of the South American continent with their longitudinal axis directed about northwest-southeast (Fig. 1).

All islands belong to the Kingdom of the Netherlands. Within this constitutional context two autonomous Caribbean "landen" (countries) are recognized: Aruba ("status aparte" since 1986) and the Netherlands Antilles. The Netherlands Antilles consist of the southern (Leeward) Caribbean islands Curaçao and Bonaire, and the Northern (Windward) Caribbean islands St. Maarten, St. Eustatius and Saba. Aruba has a complete autonomy regarding its internal affairs. When in the following chapters the term "Leeward Netherlands Antilles" (or – as a synonym – ABC-islands) is used including Aruba, this is done for reasons of simplicity only and no political or constitutional meaning is intended.

Aruba is the smallest and most western island, situated just north of the Venezuelan peninsula of Paraguana (Fig. 2). Nevertheless, it has with 360 persons/km² the highest population density of the islands. Besides that, of all islands it has the most developed touristic infrastructure with more than 720,000 tourists add to the local population of 71,000 during the year 2000.

Bonaire is situated approximately 100 km north of Venezuela (Fig. 3). The island is 40 km long by 11 km at its widest point, with a total surface area of 288 km².

The small uninhabited satellite island of Klein Bonaire is located 750 m off the western coastline of Bonaire; it is privately owned and entirely undeveloped. With a population density of only 49 persons/km², Bonaire is the least densely populated island of the Netherlands Antilles. The main center of population, Kralendijk is located in the center on the south coast of the island; a second and older center, Rincon, is located in the north. Nearly 20% of the total land area of Bonaire has been protected as a national park and since 1979, the waters around Bonaire, from the high water mark to the 200 foot depth contour, have been designated a marine park.

Curaçao, the largest island, covers an area of 444 km² and has a population of about 150,000 inhabitants (Fig. 4). Willemstad, the largest city on the island, is the capital of the Netherlands Antilles.

Klein-Curaçao, located about 10 km SE of Curaçao, is only 2.5 km long and 700 m wide. The small island is like Klein-Bonaire uninhabited besides some local fishermen living there sporadically.

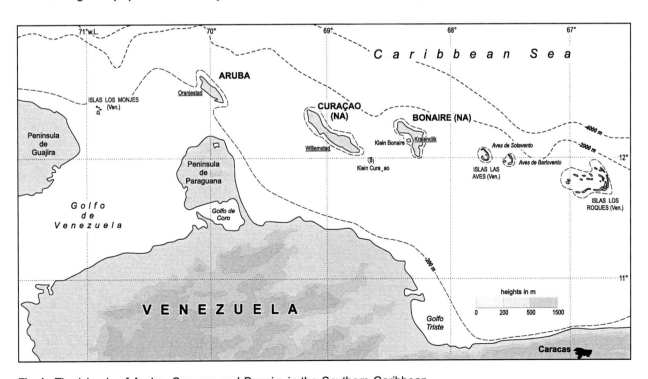

Fig. 1: The islands of Aruba, Curaçao and Bonaire in the Southern Caribbean.

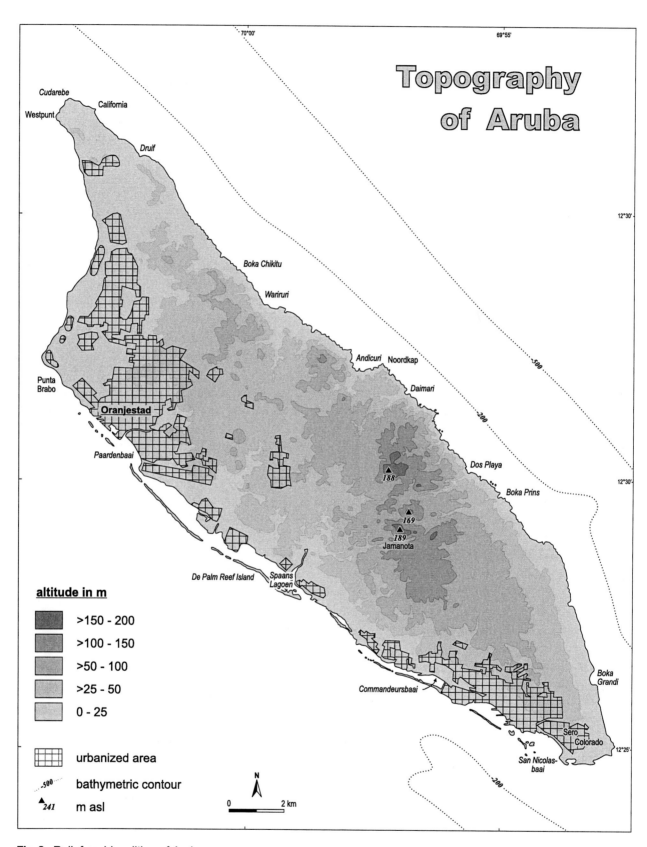

Fig. 2: Relief and localities of Aruba.

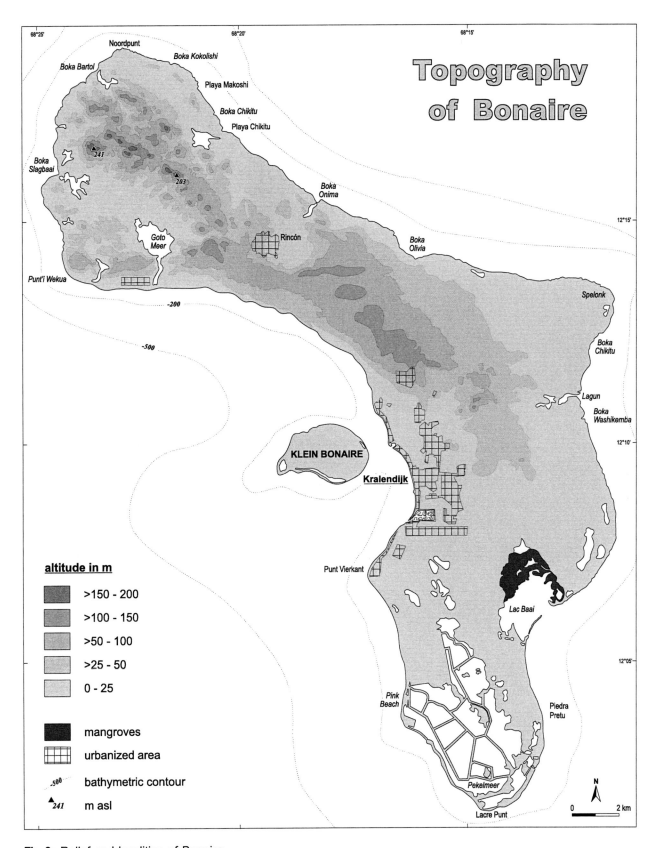

Fig. 3: Relief and localities of Bonaire.

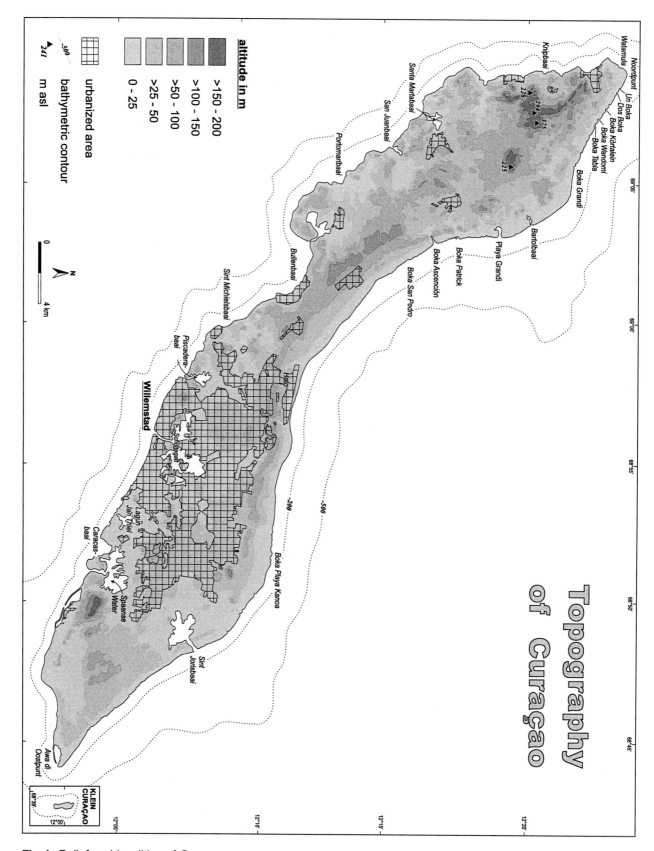

Fig. 4: Relief and localities of Curaçao.

Tab. 1: Geographical data on Aruba, Curaçao and Bonaire (DE JONG, 1999).

	Aruba	Curaçao	Bonaire
surface area (km^2)	193	444	288
max. length (km)	30	64	40
max. width (km)	5.5	16	12.5
highest altitude (m)	189	374	241
population	87,971	152,700	14,200
population density (per km^2)	440	341	49
capital	Oranjestad	Willemstad	Kralendijk
inhabitants of capital	16,000	43,600	2,000

2.2 History

A detailed overview about the "Indian period" is given by BOERSTRA (1982) and HAVISER & JAY (1987). About the first inhabitants of the ABC-islands, the Amerindian Arawaks, is not much known, most probably they migrated from Venezuela 2500 BP. Around 500 BP a new group of Indians from Venezuela – the Caiquetos – arrived at the islands. When the Spaniards under the soldier Alonso de Ojeda, joined by Amerigo Vespucci, discovered the Leeward Antilles in 1499, approximately 2000 Caiquetos lived on the ABC-islands. The dry climate, which restricted agriculture and the absence of valuable mineral resources, made the islands not attractive for the Spaniards. In 1513 the Spaniards declared the islands as "Islas Inútiles" and they were opened up for slave traders. Finally, the Dutch West India Company claimed the islands in 1634. The company installed the Dutch explorer Peter Stuyvesant as governor in 1642, and he soon established plantations on the island.

The following centuries were characterized by an increase in agriculture and the clearing of land and cutting and export of wood (HARTOG 1968, TEENSTRA 1836-1837, TERPSTRA 1948). On the plantations of the south coast of Curaçao and Bonaire saltpans were exploited. Although the plantations were not of much use for the Netherlands, they have put a stamp upon the present landscape of the islands. After the emancipation of the slaves in 1863, the plantations were abandoned and many of them were sold, mostly to the government.

During the 19th century, guano and phosphate deposits were exploited mainly on Curaçao, Klein-Curaçao and Aruba, where also some gold reserves were found. In 1920, oil was discovered off the Venezuelan coast. This signaled a new era for the ABC-islands. Aruba and Curaçao became centers for distilling crude oil imported from Venezuela, and Curaçao's Royal Dutch Shell Refinery became the island's most important business company. Nowadays the islands are dependent on subsidence from the Netherlands, oil refinery, offshore banking, and tourism.

2.3 Climate

The climatologic data of the islands are collected by the METEOROLOGICAL SERVICE OF THE NETHERLANDS ANTILLES AND ARUBA (2001) and published for the time-period from 1971-2000 for Curaçao (Tab. 2) and from 1951-1980 for Bonaire and Aruba. In general, the climate is semi-arid to arid (Fig. 5).

The average annual precipitation in Curaçao is 553.4 mm/year with about half falling in October

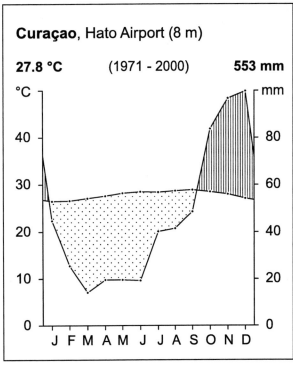

Fig. 5: Temperatures and precipitation on Curaçao (Hato Airport, 8 m asl).

Tab. 2: Climate data of Curaçao for the years 1971-2000, Hato Airport.

Curaçao, Hato Airport (1971-2000)

Element	Unit	Jan	Feb	Mar	Apr	May	Jun	Jul	Aug	Sep	Oct	Nov	Dec	Year
Avg. Air Temperature	°C	26.5	26.6	27.1	27.6	28.2	28.5	28.4	28.7	28.9	28.5	28.0	27.1	27.8
Avg. Maximum Temperature	°C	29.7	30.0	30.5	31.1	31.6	32.0	31.9	32.4	32.6	31.9	31.1	30.1	31.2
Abs. Maximum Temperature	°C	32.8	33.2	33.0	34.7	35.8	37.5	35.0	37.4	38.3	36.0	35.6	33.3	38.3
Avg. Minimum Temperature	°C	24.3	24.4	24.8	25.5	26.3	26.4	26.1	26.3	26.5	26.2	25.6	24.8	25.6
Abs. Minimum Temperature	°C	20.3	20.6	21.0	22.0	21.6	22.6	22.4	21.3	21.7	21.9	22.2	21.1	20.3
Avg. Seawater Temperature	°C	26.0	25.9	25.9	26.5	26.8	27.2	27.4	27.7	28.2	28.2	27.9	26.9	27.1
Avg. Air Pressure (-1000)	hPa	13.4	13.5	12.9	12.1	12.0	12.9	13.4	12.5	11.5	10.9	11.1	12.7	12.4
Avg. Relative Humidity	%	77.4	76.7	76.1	77.2	77.2	77.1	77.8	77.3	77.5	79.0	79.6	78.9	77.7
Avg. Daily Evaporation	mm	5.8	6.5	7.4	8.0	8.0	8.3	7.9	8.3	7.8	6.5	5.3	5.2	7.1
Highest Rainfall in 24 hours	mm	34.6	36.4	40.4	69.0	77.7	37.4	69.2	94.7	70.4	105.8	117.8	95.4	117.8
Avg. Days w. Rain >= 1.0 mm	days	8.6	5.8	2.8	2.8	2.0	3.0	6.4	5.1	4.6	7.4	9.9	11.5	70.4
Avg. Hours with Rainfall	hours	77.3	48.6	33.0	32.5	33.1	36.3	45.3	41.2	38.4	55.9	80.0	98.5	620.0
Avg. Monthly Rainfall	mm	44.7	25.5	14.2	19.6	19.6	19.3	40.2	41.5	48.6	83.7	96.7	99.8	553.4
Avg. Cloud Coverage	%	37.6	39.1	41.3	49.6	52.2	51.0	46.6	44.7	48.2	50.8	48.3	43.7	46.1
Avg. Daily Sunshine Duration	%	72.8	74.6	72.5	66.2	65.9	69.5	72.6	76.5	70.5	66.8	68.0	68.1	70.3
Avg. Global Radiation	MJ/m⁻²	151.6	147.2	179.1	168.1	183.4	176.1	193.4	187.8	179.5	161.2	133.5	136.6	1997.5
Avg. Wind Direction	degrees	86.0	85.0	85.0	85.0	89.0	91.0	89.0	89.0	88.0	88.0	84.0	83.0	87.0
Avg. Wind Speed (at +10 m)	m/sec	6.8	7.0	7.0	6.9	7.0	7.5	7.0	6.7	6.2	5.6	5.8	6.4	6.6
Avg. Wind Energy Potential	kWh/m⁻²	161.0	155.4	171.9	171.9	175.8	189.6	181.2	158.2	124.4	102.5	103.9	135.1	1830.9
Avg. Maximum Wind Speed	m/sec	12.7	12.5	12.2	12.3	12.9	13.8	13.5	12.7	11.9	11.3	11.5	12.4	12.5
Strongest Gust	m/sec	19.5	19.5	20.0	19.5	19.0	20.0	25.7	21.1	24.7	22.1	22.6	23.1	25.7
Persistency of the Wind	%	95.9	97.4	97.9	95.6	97.7	98.1	97.8	95.7	93.4	93.0	91.5	96.6	95.9

Fig. 6:
Nearly all exposed vegetation on the islands of Aruba, Curaçao and Bonaire (except of the cacti) show a significant adaptation in their growth form to the strong and persistent easterly trade winds. Divi-divi tree at Boka Washikemba, Bonaire.

to December. Bonaire and Aruba receive slightly less rainfall with a long-term average on Bonaire of 490.5 mm/year and even less on Aruba with 425.5 mm/year. As on Curaçao more than 50% of the annual rainfall take place in the last quarter of the year. Only in Bonaire, during November the average monthly rainfall exceeds 100 mm, the critical point where evaporation exceeds precipitation in tropical areas. The average temperatures on the islands are 27.5°C (Curaçao) and 27.8°C on Aruba and Bonaire, with only small monthly variations throughout the year. The coolest months are January and February with temperatures around 26.5°C and in the hottest month, September, the average temperature reaches 28.8°C. Nevertheless, there is a large variability in between the years and over each island.

In general, the eastern parts of the islands are characterized by less rainfall than the western parts with their hilly topography.

A characteristic climatological feature of all islands is the strong and persistent tradewind, which scourges the eastern and northeastern coastlines. Consequently the south, southwest and west coasts are more sheltered. The tradewind has an annual average speed of 6.6 m/sec (24 km/h) and an easterly direction (Curaçao: 87 degrees; Bonaire 79 degrees; Aruba 88 degrees) throughout the year.

The percentage of the annual persistency of the wind reaches over 95% for all islands. The mean monthly wind velocity shows a drop in the month September, October, November and December. Calm days occur occasionally in April and May. Superimposed is a diurnal cycle, the wind gathers in strength during the morning and slows down in the late afternoon. Wind modeled growth forms of the vegetation illustrate the persistency of the trade winds (Fig. 6). The strongest wind gusts have an average velocity of 25.7 m/s (92.5 km/h), which is relatively low compared to islands like e.g. Sint Maarten, situated within the hurricane belt, where the strongest gusts with values of 185 km/h illustrate the presence of frequent tropical cyclones.

The Leeward Netherlands Antilles are lying on the southern fringes of the hurricane belt. On the average once every 4 years tropical storms pass north of the islands within 100 nautical miles (METEOROLOGICAL SERVICE OF THE NETHERLANDS ANTILLES AND ARUBA, 1998). Roughly once every 100 years tropical storms passes over or just south of the islands. The hurricane climatology as well as the historical hurricane record together with the associated geomorphologic evidences are discussed in detail in Chapter 3.

The average annual seawater temperature reaches about 27°C. The mean tidal range is about 30 cm (maximum 55 cm) with a superimposed yearly oscillation of about 15 cm of the mean tide level (DE HAAN & ZANEVELD, 1959). Within the marine environment the general direction of the main current is reliably predictable in the trade wind region. Along the leeward and windward coasts of the islands a northwest current usually runs with a velocity rarely exceeding 1 knot (VAN DUYL, 1985). Swell is commonly present along the leeward sides of the northern tips of Curaçao and Bonaire. This refracted water movement originates from oceanic swell generated by the enduring fetch of the tradewinds. Its wave crests advance southward, perpendicular to the coastline and dissipate in about 10 km from the northern point.

In general, the highest wave energy environments with waves of 2.0-3.5 m height can be found along

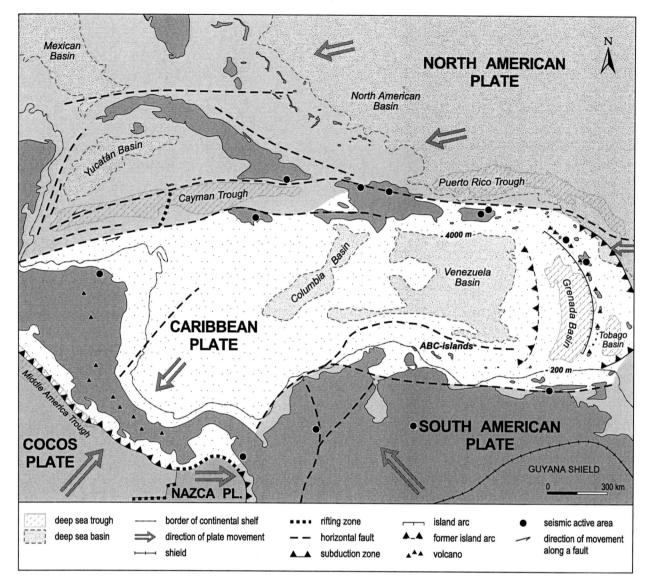

Fig. 7: Main structural features and geodynamic situation in the Caribbean: Plate boundaries, transform faults, subduction zones and volcanic island arcs (mod. after SCHUBERT, 1988; MANN et al., 1990).

the windward coasts, whereas the leeward coasts show a more complex pattern depending on the exposure of the coastline with the lowest wave heights (0-30 cm) within the sheltered bays (e.g. Daaibooibaai, Portomariebaai, Boka Santa Cruz on Curaçao) and the inland bays.

2.4 Geodynamics and Bathymetry on a Regional and Caribbean-wide Scale

Curaçao, Aruba and Bonaire are situated within the South American-Caribbean Plate Boundary Zone, along which the Caribbean plate moves eastward or relative to South America in a westward direction. The islands belong to the Aruba-La Blanquilla Chain, an E-W-running row of small islands and atolls on the Venezuelan continental borderland (also referred to as the Leeward Antilles arc). The island chain includes the Leeward Netherlands Antilles and the Venezuelan Antilles (Las Aves, Los Roques, La Orchila and La Blanquilla) (Fig. 7).

The Caribbean-South American plate boundary zone has been interpreted as a transfer zone connecting the Lesser Antilles subduction zone in the east to the subduction zone near Colombia in the west (MACDONALD et al., 2000; PINDALL & BARRET, 1990; SYKES et al., 1982). It is suggested that the boundary zone between the Caribbean and South American plates is a zone of right-oblique convergence. This situation is unique in that the Leeward Antilles arc has been colliding since the Miocene in eastern Venezuela. As a result the Venezuelan continental borderland is a tectonically complex area of depositional basins and submarine ridges that are dissected by large fault systems (MALONEY, 1967).

The ABC-islands lie within a region called the Bonaire Block, which is delimited by the Southern Caribbean Deformation Belt (Curaçao Ridge) to the north and west, the La Orchila Basin to the east and the Oca-Cuiza-San Sebastian Fault zone to the south (BEETS et al. 1984, MUESSIG 1984, MANN et al. 1990). Curaçao and each of the other islands are the emergent portions of a thick submarine ridge composed mostly of basalts and diabas. The islands are separated by faults that strike parallel to the length of the islands, creating a linear series of horsts separated by 1 to 2 km thick sediment filled submarine canyons (FOUKE, 1994; MALONEY, 1967; MUESSIG, 1984).

Curaçao and Bonaire are separated from the Venezuelan continent by the Bonaire Basin (depth >1000 m), contrary to Aruba, which is part of the Venezuelan continental shelf (Fig. 8).

The distance between Aruba to the Venezuelan mainland is about 35 km with a maximal water depth of 190 m. The Bonaire Basin itself has a flat bottom, the slope on the south is of low angle (2-4°) whereas the slope to the north is less uniform with an average inclination between 2-5° (MALONEY, 1967). The steepest portions of this slope occur in between -400 to -1100 m opposite the islands. The continental slope north of the Aruba-La Blanquilla Chain has its greatest inclination east of Bonaire where it drops down from 1829 m to 4575 m in a distance of only 12 km. The basis of the slope lies at a depth of 4865 m in the Los Roques Trough. West of Bonaire, the continental slope is less abrupt and less deep. The Caribbean Basin is separated by the Aves Swell into the Venezuelan Basin (depth >4000 m) located to the west and the Grenada Trough to the east.

The complex regional tectonic setting is the expression of the plate tectonic history in the wider Caribbean Region. Although details of the process are still highly controversial, it is generally accepted that the islands were situated on the leading edge of the oceanic Caribbean plate moving into the Caribbean region from the west Pacific Ocean (MACDONALD et al., 2000; PINDALL & BARRET, 1990; SYKES et al., 1982).

At the end of the Cretaceous, this plate collided with parts of the northern margin of the South American continent (PINDELL & BARRETT, 1990; AUDEMARD, 2001). This collision and the subsequent uplift in the Early Tertiary has largely been accommodated by folding, faulting and vertical displacements along the Southern Caribbean Deformation Belt (MUESSIG, 1984; BEETS, 1972; KLAVER, 1987). FOUKE (1994) showed that substantial vertical displacements took place also in the Middle Miocene. The ensuing Plio-Pleistocene is a tectonically quiescent period of gradual uplift based on the measurements and dating of planation surfaces and limestone terraces (SCHUBERT & VALASTRO, 1976; HERWEIJER & FOCKE, 1978; SCHUBERT & SZABO, 1978; MANN et al., 1990; RADTKE et al., 2002).

The northeastern and eastern boundary of the Caribbean Plate is formed by the West Indies island arc, which extends east from the Dominican Republic to the Lesser Antilles and then south to Trinidad (see Fig. 7). The volcanic island arc is the expression of these westward movement and the subduction of the Atlantic plate under the Caribbean plate. The portion of the West Indies arc near the Lesser Antilles (Saba to Grenada) is noted for its high volcanism during historic times. Pleistocene-Recent volcanoes occur in narrow zones less than 10 km wide with spacing between the active volcanoes varying between 15-125 km (MACDONALD et al., 2000). During the past 0.1 Ma thirteen volcanoes are believed to have been active - three are known to erupt in historic times (MACDONALD et al. 2000; TOMBLIN, 1981).

The most obvious feature of the Caribbean plate boundary is the concentration of seismic activity that almost encompasses the entire Caribbean circumference and the nearly absence of activity within the Caribbean Basin itself. The northern boundary of the Caribbean Plate is a strike-slip fault system, which crosses southern Cuba and ends in the eastern Dominican Republic, where the frontal subduction of the Lesser Antilles begins. The seismic activity increases near Puerto Rico and the Dominican Republic. In this region the depth of earthquakes increases both from north to south and from east to west; the deepest shocks occurring west of Puerto Rico (MOLNAR & SYKES, 1969). ROBSON (1964) published an earthquake catalogue for the Eastern Caribbean in which he listed 14 earthquakes in a 255 time-period (1530-1960) – most of them located in the Dominican Republic and Haiti. A regionalization of earthquake sources related to reported tsunamis shows that most events are related with both, the Northern as well as the Southern American-Caribbean Plate Boundary Zone.

The high risk of tsunami hazards for the Caribbean illustrates an empirical estimation by FERNANDEZ et al. (2000): 100% of large earthquakes (Ms 7.0) along the Caribbean plate boundary have generated tsunamis.

Fig. 8 *(see next page):*
Bathymetric map of the Caribbean Sea: A deep-sea basin of -4000 m to -5000 m forms most of the area. The eastern border is formed by the Lesser Antilles island arc system, which is the expression of the subduction of the Atlantic plate beneath the Caribbean plate. Most of the active volcanoes are located in the central part of the arc. The islands of Curaçao and Bonaire rise from a depth of more than 2000 m. They are separated from the South American mainland by the Bonaire basin, while Aruba is situated on the Venezuelan shelf.

PALEOTSUNAMIS IN THE CARIBBEAN

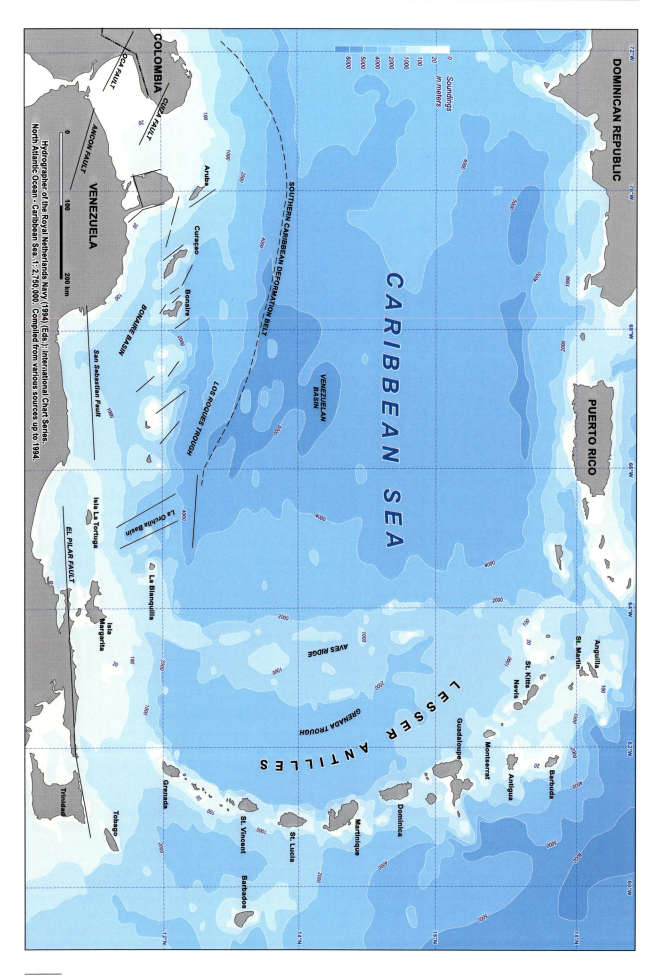

2.5 Geology and Geomorphology

Simplified, the islands consist roughly of an old cretaceous core made of metamorphosed volcanic rocks nearly entirely surrounded by Neogene and Pleistocene limestone deposits (Figs. 9-11).

Cretaceous and Early Tertiary History
The oldest geologic formations are characterized by strong submarine volcanism during the Early Cretaceous (Fig. 12). The basalts often show pillow structure, indicating submarine extrusion. The extrusion of lava is suggested to be a very rapid event which built up a basaltic layer of approximately 5000 m thickness (KLAVER, 1987). Two lavas from Curaçao have been dated by $^{40}Ar/^{39}Ar$ techniques at 88-90 Ma (SINTON et al., 1993; SINTON et al., 1998), indicating that the bulk of the lava pile was built over a short period (1-2 m.y.). At the end of the phase this layer was emerged to a certain extent, which is documented by terrigenous detritus, which is all volcaniclastic debris.

The Curaçao Lava Formation, the Washikemba Formation of Bonaire and the Aruba Lava Formation have been formed in that phase (BEETS & MAC GILLAVRY, 1977; KLAVER, 1987). On Aruba, the volcanic basement occurs mainly in the central and northeastern part of the island in a hilly landscape. Here, the highest points of the island like the Jamanota (188 m) and the Arikok (185 m) are situated. The major part of the Aruba Lava Formation has been metamorphosed by the intrusion of a ~85 to ~82 Ma tonalite gabbro batholit, called the "Aruba Batholit" (WHITE et al., 1999). A wide variety of rock differentiates can be distinguished (WESTERMANN, 1932). The rocks crop out over large areas as roundish and exfoliated monoliths, which lie in plains of diorite detritus as a consequence of former deep chemical weathering due to a more humid climate (HELMERS & BEETS, 1977). Steep hooibergite hills are encountered in several places in the diorite landscape, like the Hooiberg (167 m), a national symbol of Aruba.

On Bonaire the main part of the unit is situated in the north (National Park Washington Slagbaai), where a gently northeastern dipping section is exposed between Wecúa in the south and Salina Matijs in the north (BEETS, MAC GILLAVRY & KLAVER, 1977).

The Curaçao Lava Formation is exposed in the cores of the two anticlinoria in the southeastern and northwestern part of the island, and in a third smaller outcrop in the extreme northwest, in the Shete Boka Park area (BEETS, 1972). Stratigraphically, the Curaçao Lava Formation is followed by the Knip Group of Late Cretaceous Age. They are separated by an unconformity. Before the Knip Group was deposited, the top of the basalts was weathered, which is indicated by in situ breccias. Important evidence for the conditions during this interval is the occurrence of shallow-water limestones: the Zevenbergen limestone lenses in the northwestern part and the Casabao limestone lenses in the central part of the island. The Knip Group consists almost exclusively of pelagic and clastic, silica-rich sediments. The formation has its most extensive development in the northwestern part of the island and forms the highest elevation of Curaçao – the Christoffel Mountain (374 m). The following Midden Curaçao Formation is a conformable succession of Maastrichtian to Paleocene age, which is exposed in the synclinorium of the central part of Curaçao, and on the northern flank of the anticlinorium of southeast Curaçao. Eocene formations occur mainly in Curaçao and Bonaire. On Curaçao they are limited to four scattered outcrops of limestone, marl and clay, the thickest series of sediments is found at the Ser'i De Cueba. The Eocene on Bonaire is found around the village of Rincon and southward to Pos Dominica and with some outcrops along the windward coast.

Neogene to Quaternary History
Early Miocene mudstones and sandstones are encountered in a borehole in the center of Oranjestad, but are nowhere exposed at the surface. The tectonic uplift, which was initiated in the Middle Miocene, brought the upper surface of the islands basement repeatedly into shallow marine photic zone environments, which triggered carbonate production and the resulting Neogene deposits.

The Neogene Seroe Domi
The Neogene Seroe Domi Formation is a 350 m-thick sequence of mixed coralgal limestones and siliciclastic sandstones, which is mainly found along the leeward sides of the islands as inclined ridges onlapped by the Quaternary Limestone Terraces. The angle of the dip varies between 5° and 30° seaward. The inclined dip and leeward position was interpreted as an original leeward depositional slope in deep forereef marine environments based on: 1) in situ *Lithophagus geopetalis*; 2) strike and dip parallel to present-day shelf and coastline; 3) present-day easterly wind, wave and current motion; 4) poorly sorted sediments with no in situ corals; and mixed shallow and deep water faunal assemblages (DE BUISONJÉ, 1974). FOUKE (1994) studied the deposition, diagenesis and dolomitization history in detail. A characteristic outcrop of the Seroe Domi Formation can be found immediately northwest of Willemstad, with the so called "three brothers": Veerisberg (138 m), Jack Evertszberg (115 m) and Seru Pretu (138 m).

The Quaternary Limestone Terraces
During the Pleistocene, significant reef development occurred on all sides of the islands mostly in the

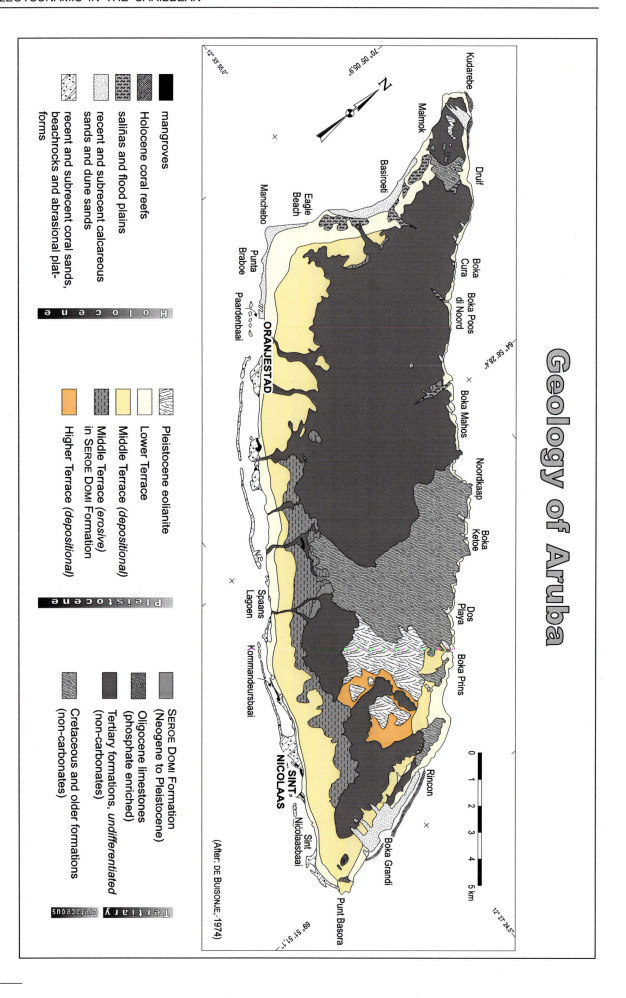

Fig. 9: Geology of Aruba: Pleistocene fringing reefs of at least three different isotope stages surrounding a non-carbonate volcanic basement (mod. after DE BUISONJÉ, 1974)

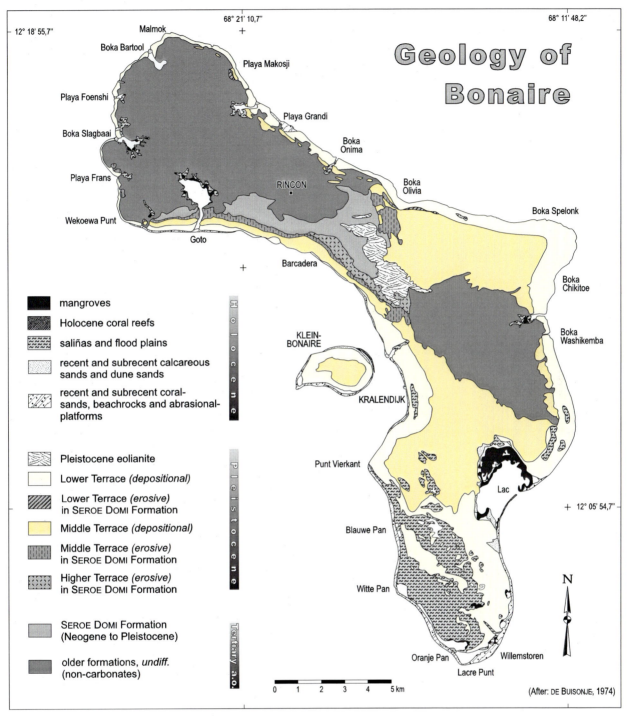

Fig. 10: Geology of Bonaire: Pleistocene fringing reefs of at least three different isotope stages surrounding a non-carbonate volcanic basement (mod. after DE BUISONJÉ, 1974).

form of fringing reefs (see Figs. 9-11). Global sealevel changes, coupled with regional tectonic uplift of the islands, resulted in the formation of five raised terrace horizons, composed of fossil reefs, which extend over long distances along the coastline of the islands. The lithofacies, stratigraphy and radiometric ages of the terraces have been described by ALEXANDER (1961), DE BUISONJÉ (1974), BANDOIAN & MURRAY (1974), HERWEIJER & FOCKE (1978), SCHUBERT & SZABO (1978) and RADTKE et al. (2002).

The lowest fossil limestone deposit will be described in more detail than the older and higher units, in order that this terrace is the basal surface for the studied hurricane and tsunami deposits. The terraces contain a high-diversity coral assemblage dominated by *Montastrea annularis*, *Acropora palmata*, *Siderastrea sidera* and *Porites porites*, as well as sands composed of fragments of coralline algae, molluscs, benthic foraminifers, echinoderms, pelagic foraminifers and siliciclastic sands and gravels. DE

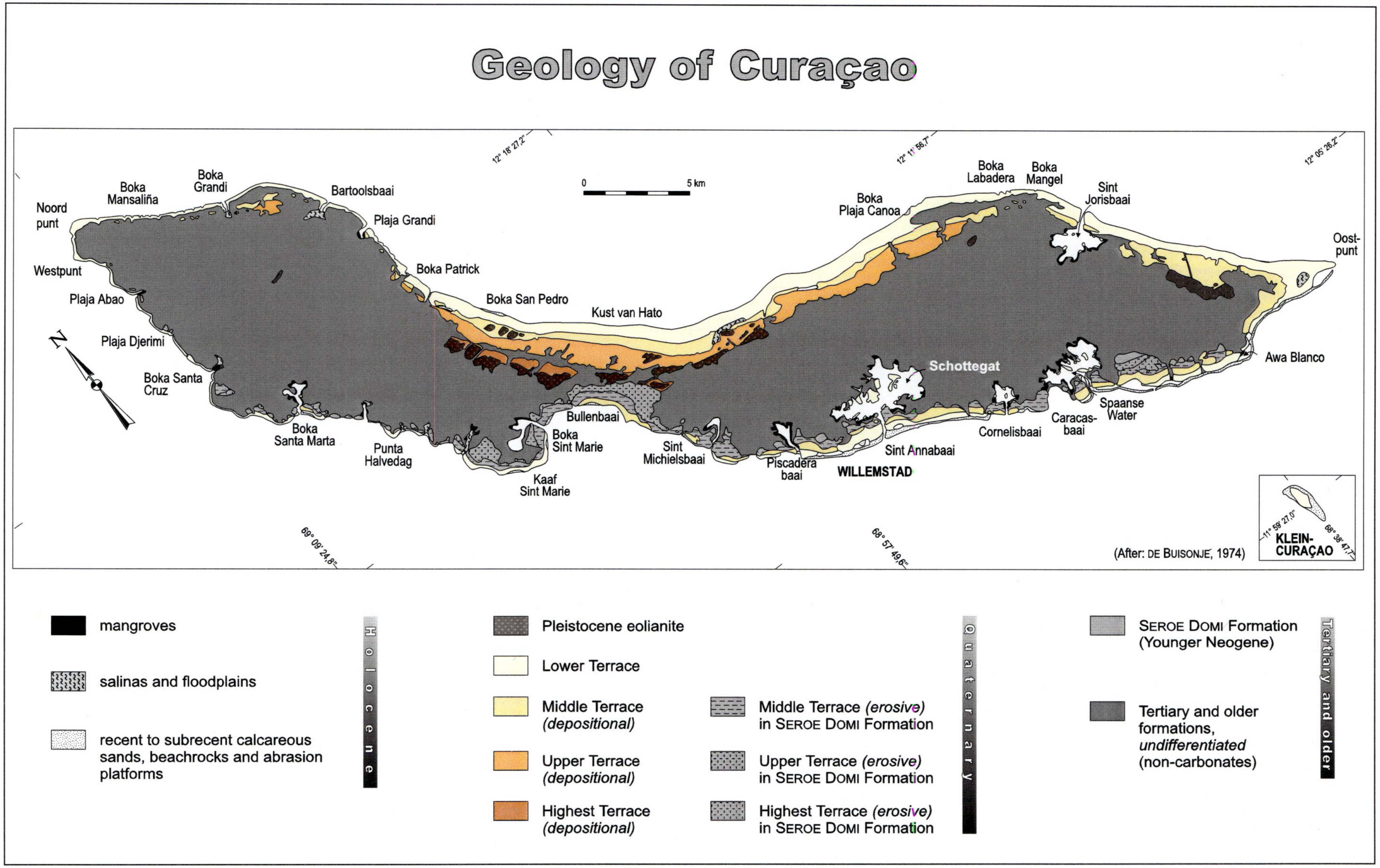

Fig. 11: Geology of Curaçao: Pleistocene fringing reefs of at least three different isotope stages surrounding a non-carbonate volcanic basement.

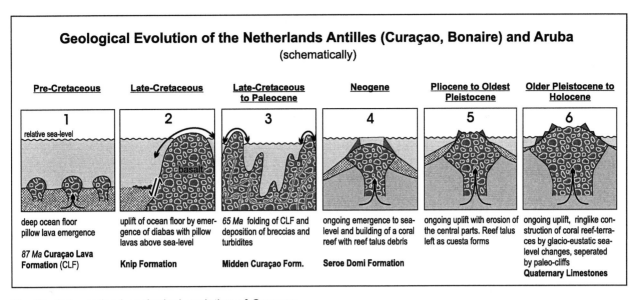

Fig. 12: Schematized geological evolution of Curaçao.

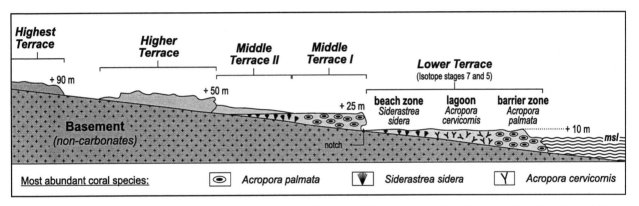

Fig. 13: Geological cross-section of the modern island of Curaçao. The cross-section illustrates the elevation of depositional and erosional terraces of the Seroe Domi Formation and Quaternary Limestone Terraces (= fossil coral reefs) (DE JONG, 2000).

BUISONJÉ (1974) subdivided the terraces into three facies, which in a landward to seaward progression include: 1. *Siderastrea* Zone (lagoonal);
    2. *Montastrea annularis* Zone (lagoonal);
    3. *Acropora palmata* Zone (fringing reef).

DE BUISONJÉ (1974) originally described five distinct horizons or steps within the Curaçao terrace sequence and interpreted them to represent in situ carbonate deposition in shallow-water shelf and reef-front environments during successive highstands of Pleistocene sealevel. According to their relative elevation, DE BUISONJÉ (1974) named the terraces as: Highest Terrace, Higher Terrace, Middle Terrace II, Middle Terrace I and Lower Terrace (Fig. 13). Steep cliffs separate the terraces, sometimes with deep incised notches at their base (Fig. 14).

Presumed lithologic equivalents of these major terrace sequences were then described and mapped on La Orchila (SCHUBERT & VALASTRO, 1976) and La Blanquilla (SCHUBERT & SZABO, 1978). Studies by BANDOIAN & MURRAY (1974) and HERWEIJER & FOCKE (1978) on Curaçao, Bonaire and Aruba have determined that each of these stepped terrace horizons is actually a complex stacked sequence of two or more depositional and erosional events that often are laterally discontinuous. HERWEIJER & FOCKE (1978) have described at least 10 depositional subunits, each of these subunits is capped with truncation surfaces and caliches depleted in $\delta^{13}C$. At least three submerged terraces have been observed on Bonaire (HERWEIJER & FOCKE, 1978). At the windward coast of Curaçao at least two submarine terraces occur (FOCKE, 1978b). The Highest and the Higher Terraces have not been radiometrically dated, but the coral composition indicates a Plio-Pleistocene age of 2.5-1.8 Ma (PANDOLFI et al., 1999). The Middle Terrace has ages between 410-500 ka, which correspond with the sealevel maxima from stage 11 in deep sea stratigraphy (HERWEIJER & FOCKE, 1978; SCHUBERT & SZABO, 1978). The uppermost unit of the Lower Terrace (Hato Unit) with an elevation

Fig. 14:
Paleo-cliff undercut at about 10 m asl by a deep bioerosive notch formed during the sealevel highstand of the last interglacial (isotopic stage 5e) in the Middle Terrace near Hato Airport, Curaçao.

between 8-12 m asl, is correlated with the 125 ka sealevel interval stage 5e (HERWEIJER & FOCKE, 1978), and for the underlying Cortalein Unit an age of 180- 25 ka (stage 7) is suggested.

A recent investigation to determine the age of the Limestone Terraces with the ESR-dating technique by RADTKE et al. (2002) suggests an age for the Lower Terrace between 75,000 and 125,000 ka reflecting at least the sealevel highstands of isotope stages 5a-e.

Each of these studies attempted to estimate the tectonic uplift rate of the island archipelago. The results yield to low uplift rates that range from 0.02 to 0.08 m/1000 years, with an estimated average of 0.05 m/1000 years (HERWEIJER & FOCKE, 1978). The height of the Lower Terrace in combination with the ESR-ages suggests a Pleistocene tectonic uplift rate at the upper end of the values, with an average uplift of 0.1 m/1000 yr. Assuming a continued and consistent slow uplift rate in the Pleistocene and Holocene, it can be concluded that the Cortalein Unit (isotope stage 7) correlates with a lower sealevel event. Isotope stage 9, so far known, has no corresponding record in the terrace sequence on land.

The Holocene sealevel reaches the present high stand in the Southern Caribbean approximately 5000 years BP (RULL, 2000), which leads to a maximum tectonic uplift of 0.6 m with an average uplift of 0.1 m/1000 years for this 6000-year time period (Fig. 15). The stability of the recent sealevel high stand is reflected in a variety of coastal morphologic features like "notches" and "benches" (see Chapter 2.5.1).

The facies of the Pleistocene deposits indicate that highest energy conditions occurred in a barrier reef along the present-day windward coast and lower energy conditions occurred along the present-day leeward coast. The orientation of the facies is parallel to the modern coastline all around the island. PANDOLFI et al. (1999) conclude from this, that on Curaçao, the relative position of windward and leeward coasts, as well as circulation patterns and tradewind direction, have not changed since the last interglacial high stand 125 ka years ago.

The height of the top-surface of the old reef crest – at many places geomorphological to distinguish from a lower lying landward lagoonal strip – is varying from 4 to 12 m above the present sealevel. Higher terrace elevations occur in the northwestern parts and on the northeastern flanks of the islands. At the southeast sides of Bonaire and Curaçao and along the southwest coast of Aruba, the altitude is locally less than 4 m asl. This may suggest a tilting along the longer axis of the island combined with an individual movement of each island. The top surface is with a slight dip of less than two degrees almost horizontal. In general, the Lower Terrace Limestones are wider along the windward coast.

In Curaçao and Bonaire they have an average width of 600 m compared to only 200 m on the leeward side. On the east side of Bonaire in the area of Bacuna-Lac-Broea Wesoe, the terrace has an unusual width of several kilometers. Aruba shows a different situation: Whereas the southwest side has a width of up to 400 m, the northeast side has a maximum width of 150 m. The two small islands Klein-Curaçao and Klein-Bonaire consisting entirely of limestone terraces. Klein-Curaçao is exclusively made of Lower Terrace limestone deposits, Klein-Bonaire has a central part consisting of Middle Terrace Limestone surrounded by the Lower Terrace deposits. Both islands are encircled by recent to subrecent hurricane and tsunami deposits.

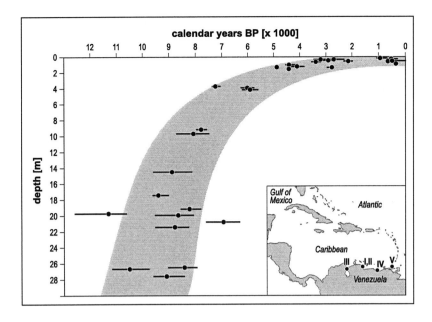

Fig. 15:
Sealevel rising trend for the Caribbean coasts of Venezuela in calendar years before present (RULL, 2000).

A subaerial effect caused by rainwater is the development of a karst surface with a topography of up to 0.5 to 1.0 m. The relief is mostly formed of small and often very sharp peaks of limestone, which protrude from the basin-like depressions lying in between. Dolines can be found near Noordpunt and Playa Kanoa, Curaçao. Development of caves caused by subterraneous limestone solution occurs mainly in higher, older limestone units (e.g. the Hato Caves on Curaçao). The caves are cut vertically in the cliff fronts between Lower and Middle Terrace and between Middle and Higher Terrace. They are mostly filled with terrigenous material like speleothems and reddish flowstones. The speleothems could be a valuable archive of the Younger Pleistocene climate history and vegetation development.

Under the present climatological and vegetation conditions the weathered materials on hill slopes are transported by sheetwash and splash erosion, which leads to the formation of gentle sloping pediments around the higher hills. In places where less resistant rocks are covered by Neogene or Quaternary limestone caps mesa- and cuesta-like forms are developed. Where these limestone caps get undermined in the foothills of cuestas or isolated table mountain, limestone blocks may break off from the cuesta scarp and gliding as fragments down the slope. Often deeply weathered diabas act as the glide basis. Field observations during this study suggest that this transport mechanism is responsible for the huge blocks situated at Caracasbaai. Evidence for a deposition by a tsunami generated by the Caracasbaai landslide as postulated by DE BUISONJÉ & ZONNEVELD (1976) could not be confirmed (see Chapter 5.2.1.1).

Furthermore geological and geomorphologic features are the recent and subrecent deposits associated with coastal processes and the overall coastal morphology, which will be described from the surface of the elevated Lower Terrace to the subsea aspects, as they are important for this study.

2.5.1 Coastal Geomorphology

A number of interacting environmental conditions, like the dry climate, the strong easterly tradewinds and the steep surrounding limestone coastlines, are the main causes for a characteristic coastal geomorphology on all islands. The most impressive contrast is the differentiation in the rough almost not accessible windward coasts with heavy surf action and characteristic geomorphologic features like wide benches and well-defined notch profiles and the leeward sides with sheltered beaches and a concentration of touristic infrastructure.

Nearly everywhere the coastline of the islands is developed in calcareous material either as cliffs or as beaches consisting of calcareous sand or coral shingle. Only for a short stretch of a few kilometers in length in Aruba near Noordkap, the volcanic basement borders on the sea. The elevated limestones are interrupted by narrow inlets and wide inland bays, the partly drowned parts of larger drainage systems during periods of lower sealevel. The latter ones – according to the term of SCHÜLKE (1969) characterized as "panfluvial rias" – occurring mainly along the leeward side of the islands with a typically lobate, hand-shaped contour (e.g. Spaanse Water, Schottegat, Sint Jorisbaai on Curaçao (see Figs. 2-4, Fig. 16).

The narrow inlets, which are bordered at both sides by steep walls (locally called 'boka", literary "mouth") are restricted to the windward coasts (Fig. 17). Many smaller bokas do not completely traverse the Lower

Fig. 16:
One of the typical ria-like embayments at the leeward coast of Curaçao (Spaanse Water).

Fig. 17:
The small, narrow embayments with steep side walls are called "boka" on the Antillean islands. They mostly appear at the windward coasts (Boka Djegu in Shete Boka Park, Curaçao).

Terrace, but are only indentations in the coastline. The larger bokas dissect the Lower Terrace and in most cases also the older limestone terraces. Often these large bokas are passages that connect inland bays with the open sea. Sometimes bars, composed of subrecent or recent coral rubble, block these passages. Examples of such closings are Lagoen Jan Thiel and St. Michielsbaai in Curaçao, and Slagbaai, Boka Bartol, and Goto Meer in Bonaire.

The largest shallow inland bay of the Netherlands Antilles, Lac Baai, is situated on the windward shore in southeastern Bonaire. The flooded area is approximately 7.5 km² with a maximum water depth within the bay of 4.5 m; tidal range is limited to approximately 0.3 m. The bay is bounded seaward by exposed fringing coral reefs, that protect the bay from wave action. Waves break over the reef, flood the bay, and flow out through a deep water channel at the northernmost tip of the bay, creating a rip current.

A scientifically and genetically correct termination of these bays and bokas is difficult in order that the genesis of the inlets is still unclear and different causes must be discussed. According to DE BUISONJÉ (1974) the narrow bokas, in cases filled up with large blocks of Lower Terrace limestones, owe their existence to a collapsed ceiling of a water drainage that previously ran under the limestone surface or are cut into the limestone by wave action. That would lead to the term "Calanquen" (GALAS, 1969; KELLETAT, 1999). If they represent as well former valley systems, the characterization as "torrentielle Kastental-Zwergrias" after the nomenclature of SCHÜLKE (1969) would be correct. Is their opening due to freshwater influence, which led to the forming of channels within the former coral reef body, they would be correctly named "Scherm" (SCHMIDT, 1923).

The discussion of the genesis is inseparably associated with considerations about the age of the bokas. Assuming they are "Scherms" they could be as old as the last two interglacial periods as they transect limestone deposits of that age. Valley inlets formed by backward incising fluvial erosion would limit their existence to the last sealevel low stand. The increasing amount of non-calcareous pebbles in the terrace limestones of larger boka transects indicate, that these boka systems already functioned as drainage basins in early Quaternary times. Whereas the origination as "Calanquen" would restrict their age to the younger Holocene with hurricane- and/or tsunami events as the main forming process orientated at primary weak construction zones within the fossil reef structure. An unambiguous classification is difficult in order that on Curaçao most bokas show all of the above mentioned genetic characteristic features.

There exist transitional forms to the classical "rias-type" (Playa Grandi) as well as initial phases of "calanquen", which are cut into the limestone deposits by marine abrasion accompanied by collapse of caves and crevices (e.g. Boka Tabla Cava, Curaçao).

Small beaches, consisting of calcareous sand and coral shingle, mixed with low and changing quantities of terrigenous material generally characterize the part where a drainage system enters the bokas. Beachrock in a destructive phase is a very common phenomenon in these beaches. Nevertheless, beaches are not exclusively confined to the boka openings, some of the larger recent beach and dune complexes extent in NW Aruba near California Lighthouse and Boka Grandi, or Boka Prins in the southeast, and in less extension on Bonaire near Playa Chikitu and Manparia Goetoe. The northwestern part of Aruba also shows sandy beaches over long stretches. Here, the beaches have developed in connection with an extremely extensive submarine platform of more than one kilometer width and a depth of only 20 m. Occurrences of eolianites in many parts of the Leeward Lesser Antilles point to the fact that in the geological past the extension of dunes was much larger than today.

Along the leeward sides of the islands long coastal sections have the character of narrow and steep beaches built of shingle derived from coral colonies growing immediately seaward of the coastline. The most common species here is *Acropora cervicornis* – a fragile coral living in less turbulent conditions. A commonly observed feature along the windward coasts are ridges formed of coarser coral rubble and boulders which are accumulated over long coastal sections in a distance of up to tens of meters of the coastline. These deposits are hitherto generally ascribed to extremely stormy weather and hurricanes.

Limestone cliffs with notches and benches genetically related to sealevel are very common in tropical areas (KELLETAT, 1999). They are the expression of bioerosive and bioconstructive processes associated with a pronounced zonation of organisms in the transition zone between the marine and terrestrial environment. This transition zone from the upper subtidal to the lower supratidal is the habitat for a variety of intertidal organisms, which are either encrusting or boring species. The zonal width of the organism distribution is depending on the extent of suitable microclimatic conditions as well as surf, splash and spray influence. The greater the tidal range or – in regions with virtually no tidal rise and fall – the greater the exposure to wave action, the further the zonal boundaries are pushed upshore, resulting eventually in a wider notch in a vertical sense. In other words, the notch morphology reveals the mean degree of wave exposition after the wave energy has been filtered through the reef, if one is present. As water turbulence increases, organic accretions appear in the middle of the notch in its most surf beaten part. These accretions become more and more pronounced and eventually form a seaward protruding bench (surf platform), so that part of the notch is above, and another part is below the bench. Thus, the rocky carbonate shorelines are predominantly sculptured by biogenous processes like bioerosion and bioconstruction.

The morphology of the cliffs of the Leeward Netherlands Antilles has been studied by MARTIN (1888), DE BUISONJÉ & ZONNEVELD (1960) and FOCKE (1977, 1978c, d). Organism distribution on the cliffs has been described by VAN DEN HOEK (1969), VAN LOENHOUD & VAN DE SANDE (1977) and FOCKE (1977, 1978c).

Along the coasts of Curaçao and Bonaire VAN DUYL (1985) distinguished nine wave energy environments. Six of them were characterized by the height of unrefracted waves generated by wind. The remaining three had comparable wave heights, but were in addition supported by refracted oceanic swell with greater wavelength. Depending on the degree of exposure to water turbulence FOCKE (1978d) described four different cliff types as examples of a continuous range of variations over the physical environmental conditions. Figure 18 shows the generalized morphology of the cliff profiles. Their spatial distribution along the coastline in dependence of the wave energy environments is illustrated in Figure 19.

The most sheltered profile shows a horizontal under-

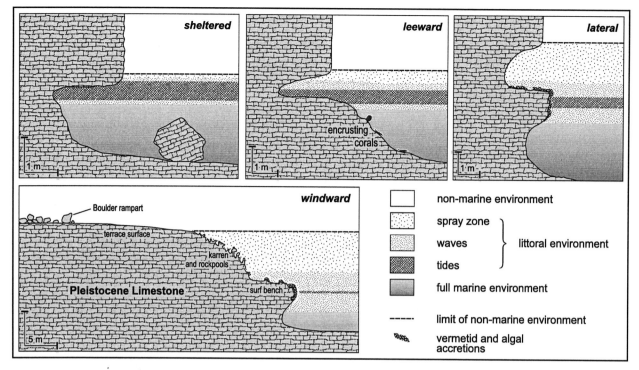

Fig. 18: Four different types of bioerosive notches on the islands of Aruba, Curaçao and Bonaire are strictly depending on the surf environment (mod. after FOCKE, 1978c).

cutting of the cliff, representing basically the difference between marine and terrestrial erosion rates. The difference is in the order of two magnitudes: Marine erosion rates are estimated at 1.0-1.5 mm/yr (HODGKIN, 1964; KELLETAT, 1999; TRUDGILL, 1976) whereas terrestrial erosion rates are estimated to be significantly slower with only 0.01-0.02 mm/yr (FOCKE, 1978d). The maximum notch depth in the sheltered environment is 3-4 m. Periodic collapse of the almost horizontal upper part is occurring sporadically if crevices are reached by the notching process. Published rates of biological destruction of coral reefs and coastal environments illustrate how few detailed studies have been undertaken. TRUDGILL (1983) has summarized much of the quantitative data available and a detailed review is given by HUTCHINSON (1986).

A common feature is the development of a new notch in the collapsed limestone blocks. The intensity of notching may reflect the relative age of the breakdown. The lower part of the notch is characterized by large numbers (>200/m^2) of the boring echinoid *Echinometra lucunter* with boring holes up to 10 cm deep, as well as sipunculid worms and boring bivalves (*Lithophaga sp.*). For a dense *Echinometra* field of approximatly 100 individuals/m^2 an erosion rate of 1.4 mm/yr is estimated (FOCKE, 1978c), this for the echinoids alone. In between the echinoid boreholes accretions with a thickness of up to 15 cm are build predominantly by the coralline algae *Porolithum pachydermum*, the foraminifer *Homotrema rubrum* and the serpulid worm *Spirobranchus polycerus var. Augeneri*.

The upper part of the notch contains virtually no accretions apart from a thin coralline algal crust, here the barnacle *Lithotrya dorsalis* and sponges predominantly bore the surface. The boreholes are 10-15 cm deep and commonly occur up to 1000 holes/m^2.

The sheltered profile is limited to low wave energy environments (waves 0-30 cm high) as they occur in the sheltered bays of Daaibooibaai or Portomariebaai and along the most sheltered stretches around Kralendijk, Bonaire and Klein-Bonaire. Along the northern coastlines of Curaçao and Bonaire this wave energy environment is supplemented with oceanic swell heading south from a northern direction. Here, and in the adjacent low to moderate wave energy environment 5 and 4 (Fig. 19) the leeward cliff profile is most common. It is characterized by a V-shaped notch-profile, which increases with increasing wave action (Fig. 20).

Where considerable oceanic swell complicates wave patterns, lateral cliff profiles are formed. Generally, the lateral profile is associated with the moderate-high wave energy environment 3. The most exposed capes and beaches are subject to this degree of wave action. Waves of 1.5 m are not exceptional. The cliff profile shows a narrow bench of 1-2 m width dividing the notch in two separate features – both are contemporaneous. The development of a bench is associated with encrusting organisms, which protect the un-

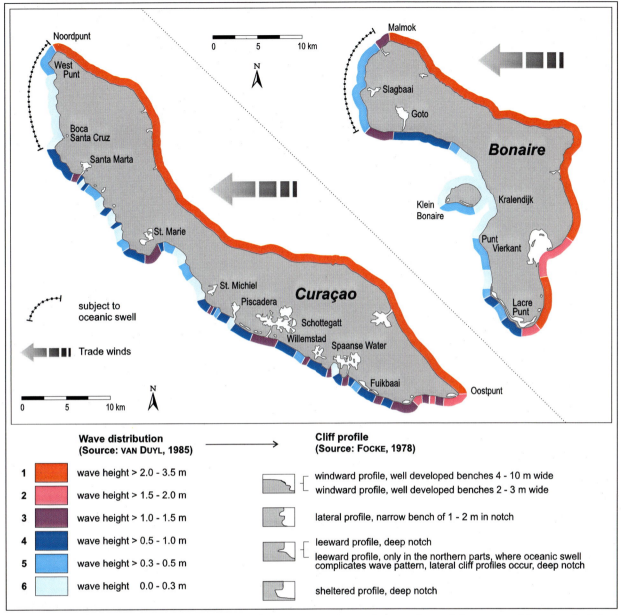

Fig. 19: The distribution of wave heights, cliff types and notches along the coastlines of Curaçao and Bonaire: A distinct difference can be seen between the windward (NE) and the leeward coasts (SW).

derlying limestone from erosion, whereas above and beneath the accretions bioerosion is the dominant process. The accretions are predominantly built by the coralline algae *Porolithum pachydermum* and *Lithophyllum congestum* and the vermetid gastropod *Spiroglyphus irregularis*. These framebuilders benefit from the increased turbulence with better nutrient supply resulting in higher growth rates. On the higher parts of the platform generally the vermetids, relative to the coralline algae, are predominant.

Comparing the areas with wave energy environments 3 and 4 between Curaçao and Bonaire it can be seen, that the leeward side of Bonaire is more sheltered than the leeward side of Curaçao. The position of Bonaire with regard to the consistent eastern tradewind offers more protection to its leeward side. The highest wave energy environments 1 and 2 are found along the windward coasts and around the southern tips of the island. The entire cliff profile is now within the marine environment. The heavy wave action is reflected by the windward cliff profile with its well-developed benches of up to 10-12 m widths and a broad supratidal spray zone with a rugged karren surface. Underneath the platform, unhindered by accretions, rapid erosion forms a notch.

All three zones – now representing a vertical interval of more than 15 m – are equivalent to the single notch of more sheltered cliffs. This cliff profile is present along most coast sections of the windward sides. On Aruba between Sero Colorado and Boka Grandi a particular spectacular bench section is developed (Fig. 21). In the southern windward part of Curaçao the benches are more narrow (3-5 m). The wave height in this area is slightly reduced due to the position of Bonaire, which results in a reduced fetch of the wind (about 46 km). The sprayzone is characterized by continuous spraying with seawater. Here the surface of the Pleistocene terrace shows an extremely rugged convex slope. The micro-topography of the surface with the protruding limestone peaks and rockpools can be as accentuated as 1.0 m, sometimes even more (Fig. 22). The sprayzone may be 15-30 m wide with a vertical distance of up to 8 m. Blue-green algae (*Cyanophyceae*) penetrate the limestones up to 1-2 mm deep (light compensation depth), which are grazed by a number of snails (notably *Littorina sp.*) forming the characteristic rockpool features (KELLETAT, 1998).

The upper surface of the bench platform is determined by the highest level at which vermetids are able to build accretions. Where exceptionally huge quantities of spraywater are flowing back to the surf platform vermetids even build accretions in the spray zone some 2 meters above mean sealevel (Fig. 23). On the surface of the bench the described framebuilders – coralline algae and vermetids – form semicircled ridges, which results in a terraced system of pools. The ridges on the seaward edge and on the surface platform form up to 70 cm thick accretions. With the surf, huge amounts of water flow over the platform. The accretions of the platform are intensively lithified. The result is a dense and very wave resistant structure. Only very rarely collapsed parts of surf platforms have been observed on the sea bottom in front of the cliffs (FOCKE, 1978b; VAN DUYL, 1985).

Once created, the presence of the platform retards the erosion rates of the spray and notch zone, resulting in an equilibrium in which the profile will not notably change as the cliff recedes landward (Fig. 24): The nearly overall distribution of tide level-depending features like notches and benches clearly indicate that the modern sealevel has been constantly existing for a long time, presumably during the total timespan of the younger Holocene, when the sealevel reached nearly its present position about 5000 years BP. This assumption does not exclude a slight regressive-transgressive oscillation, or a sealevel rise of some decimeters in these several thousands year time period. Evidences of such oscillations can be observed at one location opposite the Sea Aquarium on Curaçao. Here, over a very limited distance a short phase of notching about 0.4 m above present sealevel is represented by a small notch, which

Fig. 20: Example of a deep incised bioerosive notch at the leeward to lateral coastal section of Curaçao. No signs of sealevel variations during notch formation can be identified.

might also indicate a dislocation of this part of the island. The depth of the sea immediately in front of the cliffs varies between 3-15 m. The seabottom comprises a slightly sloping submarine terrace stretching across a distance of minimal 20 m along the leeward side to maximal 250 m on the windward coast to a 5-15 m drop-off (FOCKE, 1978b; VAN DUYL, 1985). Below the drop-off a steep slope, with a usual angle between 20-50°, runs to a depth of 25-55 m, where it gradually levels off. On Curaçao, the steepest slopes with angles >50° are concentrated in the eastern half of the island, concurrently subsea cliffs occur in the drop-off zone. Along Bonaire the steepest slopes are situated in the northern leeward part, around the southern tip and around Klein-Bonaire. The adjacent slope angle class (40-50°) confirms this distribution. Gentle slopes of less than 20° are mainly present on Curaçao along the northern leeward parts, whereas on Bonaire and Klein-Bonaire these slopes angle are rarely found.

The first shallow platform on the north coast of Curaçao has a rocky bottom devoid of unconsolidated

Fig. 21:
Between Sero Colorado and Boka Grandi at the exposed windward coastline of Aruba a very well developed bench can be seen with a width of more than 10 m. It clearly points to a long lasting sealevel during the Holocene.

Fig. 22:
Typical micro-relief in the supra-tidal spray zone on carbonate rocks along the exposed coastlines of Aruba, Curaçao and Bonaire: up to one meter deep rock pools carved by the grazing activity of gastropods (mostly littorinids), separated by very sharp edges. No coarse sediments are present in these environments.

Fig. 23
Front, surface and upper parts of the benches in the very exposed environments of the intertidal zone on Aruba, Curaçao and Bonaire often show a pattern of vermetid rims separating shallow pools. The vermetid accretions can be developed up to 2 m above mean high water and are very resistant against mechanical destruction by surf beat. They protect the underlying reef rock from bioerosion.

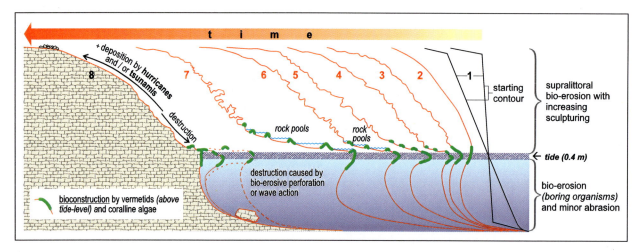

Fig. 24: The development of a bench along a steep and sediment free exposed coastline in carbonate rocks of the Antillean islands. Accretions of calcareous algae and vermetids protect a certain part of the rock in the intertidal zone against bioerosion, whereas above and below this protected zone the cliff front is cut back by bioerosion.

sediments, but densely covered with fleshy algae on which the brown macro-alga *Sargassum* is the most abundant. The absence of unconsolidated sediments on the windward coast is striking considering this environment to be the place of origin of the Holocene coral debris deposited on the shore. The second platform at depth of 32-38 m carries unconsolidated calcareous sands on which sponges, rodoliths and several species of *Strombus* are very abundant.

2.5.2 Recent Distribution of Coral Reefs

The fringing reefs along Curaçao and Bonaire are distinguished from reefs elsewhere in the Caribbean (e.g. Jamaica) in that reefs grow along leeward sides and reef development is almost absent along the windward sides at depths shallower than 12 m (BAK, 1977; FOCKE, 1978b; VAN DUYL, 1985). Whereas at the leeward coast at least 16 m of Holocene accumulation took place, nearly no extended flourishing coral communities have been found on the windward coast, where in contrast the first subsea-terrace is densely covered with *Sargassum platycarpum*, only locally scattered corals are found (*Diploria clivosa, Porites asteroides*). Reef corals like *Montastrea annularis, Acropora palmata* and *Dendrogyra cylindricus* occur mainly below the 12 m-depth contour (VAN DUYL, 1985; FOCKE, 1978b). The present-day absence of flourishing fringing reefs at the windward coasts where they are well documented in the fossil terrace sequence is, according to VAN DUYL (1985), mainly due to bottom morphology and wave exposition. VAN DUYL (1985) suggests that the tectonic uplift since the Pleistocene reduced the shelf area available for coral growth considerably. As a result the present shallow carbonate platforms (as deep as 10 m depth) with an average width of 80-150 m surrounding the islands are narrow, which emphasizes the turbulent, abrasional components of water movements. Coral survival on the shallow platform along the windward sides is likely thwarted by this vigorous wave action. Nevertheless, examples from the Pacific and Indian Ocean show that flourishing coral reef growth is occurring along very steep coasts, which are similarly exposed to strong tradewinds and hurricanes, and these reefs even have developed up to the low-water-level at many places. Concurrently the high cover of *Sargassum* must seriously limit juvenile coral survival. VAN DUYL (1985) concluded that the establishment of a critical mass of corals and coralline algae between the shoreline and the 12 m-depth contour failed to take place. Contrary, evidence from the paleo-surfaces, as described above, show clearly that the reefs during the last interglacial periods reached the former low water level – with their reef crests and lagoons well preserved till present – whereas nowadays only some spots can be found in NW Aruba, SE Aruba and SE Bonaire.

Considering that the environmental conditions like geographical position, tradewind patterns associated with similar wave height as present, did not differ significantly, it can be stated that the absence of present extend flourishing reef growth is one of the unsolved questions and still most important geo-secrets of the Leeward Netherlands Antilles.

2.6 Soils and Vegetation

2.6.1 Soils

In an early soil survey on Curaçao, (HAMILTON, 1941) the soils were described as residual soils with large mineral reserves, for a great part caused by dry physical weathering. The water deficiency on the island was not only explained as a result of the arid climate but also by the porous structure and low clay and humus content of the soils. Based on a survey for a development plan on land and water (GRONTMIJ & SOGREAH, 1968), a division was made into different soil and land types according to the geological formations and characteristics like texture and soil depth. The soils were divided into three main groups:

- soils on limestone formation;

- soils on diabases;

- alluvial and colluvial soils.

The soils on limestone formations are highly porous, generally shallow and mostly saline. They cover about 28% of the total surface area. Their drainage system is different from those in non-calcareous formations. These soils are subdivided into the reddish soils and lagoonal soils of the Lower Terrace. The light brown soils on diabases (basalt) cover about 38% of the total surface and range in thickness between 0.1-0.4 m. Alluvial and colluvial soils are found in drainage basins, locally called "roois", and plains, they are moderately deep, clayey and loamy.

In 1992 a geochemical soil survey was carried out by DE VRIES (2000), as part of an overall geochemical characterization of the Leeward Netherlands Antilles. Labelled according to their dominant geology, soil process and most striking characteristics, six dominant soils types were described: sandy limestone soil type, Midden Curaçao soil type, arid/calcareous soil type, basalt West soil type, agriculture-influenced soil type, and the basalt East soil type.

2.6.2 Vegetation

The vegetation of the islands is xerophytic. All three islands are located in the "Caribbean Dry Region", which is situated between the Araya Peninsula in Venezuela and Cartagena in Colombia (SARMIENTO, 1976). BOLDINGH (1914) described the vegetation of Curaçao as "that of a dry country, where thorny scrubs and cactuses are predominant". In the early 1950's, STOFFERS (1956) extensively studied the vegetation of the Netherlands Antilles. For the Leeward Islands he described 18 different vegetation types, but pointed out the high lack of knowledge of the extent to which vegetation had departed from the natural climax vegetation as a result of human activities. Since STOFFERS' work (1956) only few vegetation studies have been carried out. Coastal communities were studied in more detail by STOFFERS (1980) as well as limestone communities by ROJER-BEENHAKKER (1987). BEERS et al. (1997) published a landscape ecological vegetation map for Curaçao.

The vegetation on all three islands has suffered from centuries of grazing by introduced animal, mainly goats, woodcutting for export and clearing of areas for the cultivation of Aloe. Most vegetation types belong to secondary woodland and scrub vegetation. The main influencing factors for the vegetation pattern are the geologic formations and the influence of the salt spray. In general, on volcanic formations vegetation types with deciduous plant species predominate. The limestone formations are characterized by more or less evergreen plant species. Relatively simple structured mangrove formations are present along inner bays and at shallow places along the leeward coasts. The coastal vegetation of the Lower Terrace is strongly influenced by salt spray and the constant trade winds (BEERS et al., 1997). Vegetation is mainly present on the solid limestone rocky substratum and in small erosion pits with a thin clay layer. Along the northcoast, the first vegetation zone behind the "karren zone", is a very barren area with a sparse vegetation consisting of small herbs like *Lithophila muscoides*, *Paspalum laxum* and *Euphorbia sp.*. More landwards, a low but dense vegetation of windblown *Conocarpus* shrubs occurs, growing prostrate over the limestone rocks and alternating with open to almost barren spaces. The most remarkable vegetation feature is the roundly cactus *Melocactus*, which is growing preferably on the outcrops of the Curaçao Lava Formation. Further on, often an impenetrable thorny, woody vegetation, dominated by *Acacia tortuosa* trees, together with *Prosopis jul,. Randia ac.*, *Caesalpinia coriaria* (loc. Divi-divi) as well as the columnar cactus *Cereus repandus* is present (Fig. 25). In places where thicker soils have been formed *Hippomane* bushes can be found with spectacular bended forms, growing in thickets up to a length of 15 m.

Along the south coast the beaches are used mostly for recreational purposes, as a consequence there is hardly any vegetation left with the exception of some *Hippomane* trees. A few patches of still undisturbed beach vegetation can be found on almost inaccessible places, like the southeastern coast (Fuikbaai) and near Bullenbaai (BEERS et al. 1997).

Mangrove vegetation is present on the muddy to sandy and sheltered transition zones from land to sea, mostly on the shoreline of inland bays. Significant parts of the mangrove vegetation are dominated by *Rhizophora mangle*. On Curaçao, extensive mangrove areas are mainly found around the large inland bays of Spaanse Water and St. Jorisbaai. In many of the smaller bays they have been extinguished completely. The largest mangrove area of the ABC-islands is located at Lac Baai on the southeastern coast of Bonaire (Fig. 26). Along the landward edge

Fig. 25:
A typical landscape aspect of the islands. Columnar cactus (*Cereus repandus*) tower over thorn scrub vegetation.

Fig. 26:
The largest stand of mangroves (*Avicennia sp.* and *Rhizophora sp.*) on the islands can be found in Lac Baai in the southeast part of Bonaire.

of the bay an actively growing fringe of *Rhizophora mangle* with an average height of 8 m is situated. The nature of the mangrove stands is relatively rare because there is no riverine input into the system, making them particularly vulnerable. Within the mangrove system are a number of permanently dry, ribbon-like cays as well as several important feeder channels that supply water to the back of the mangrove fringe. The mangroves are dominated by *Rhizophora mangle* along the landward and seaward edges and *Avicennia germinans* on the drier grounds. Within the *Avicennia* zone, average tree height is 5 m. *Conocarpus erectus* is also common. One of the dominant features of the mangrove system is a significant die-back of *Rhizophora mangle* at its northwestern part, thought to be due to hypersaline conditions created by the landward damming of freshwater and choking of feeder channels on the seaward side resulting in water temperatures of 40°C and salinities of up to 100 ppt (VAN MOORSEL & MEIJER, 1993). Nowadays, the mangrove formations are clearly restricted by landfill, cutting for construction purposes, charcoal burning and the transformation into saltpans (Bonaire). Beside the effects of these disturbances on the spatial vegetation distribution, the mangrove vegetation is partly transformed into mangrove shrub formations. The degradation of the mangrove vegetation is therefore mainly caused by human activity in contradiction with areas in southern Florida or the Bahamas, where the disturbed mangrove vegetation can be used as an indicator for the intensity and frequency of natural disturbances by hurricane impacts (KELLETAT, pers. observ. 1978-1998). Severe Hurricanes, which can cause major damage to mangrove vegetation, are usually not passing the ABC-islands.

Fig. 27:
Saltpans cover extended areas of former lagoons and low lying Pleistocene reef flats in Bonaire.

2.7 Landuse and Infrastructure

2.7.1 Landuse

Nowadays agriculture on the islands is concentrated on Curaçao, and is limited to a few small areas mostly situated on soils of the Curaçao Lava Formation, due to the presence of aquifers. Depending on the water resource, there are two types of agriculture: The rain-dependent agriculture is limited and difficult because of water shortage most of the year. Products are mainly cucumbers, watermelons, ochra and beans. The other type is irrigated agriculture with e.g. drip-irrigation systems. The disadvantage is the effect of salt accumulation in the topsoil. Owing to this, many salt-resistant cocos palms are grown.

Salt mining plays an important role in the south of Bonaire, where the solar salt works of the Antilles Salt Cy. Ltd. hold an extensive complex of saltpans since the late sixties (Fig. 27). Seawater is being driven in the system by the trade wind through an eastern inlet cut in the coral rubble ridge. It circulates through ten condensators until it reaches a specific density and is then brought into saturating ponds where the brine stays until it reaches the saturation point of sodiumchloride. A shrimp farm started operations in 1999. STIENSTRA (1991) documents the history of phospate mining on Klein-Curaçao, Curaçao, Aruba and Bonaire. It is estimated that between 1871 and 1985 a total of at least 6 million tons of guano, phosphorite and other types of phosphate rocks were mined and exported.

Mining activities are also related to the extensive tsunami debris deposits, mostly without any legal regulations. In the past and still in the present the coral rubble are used as construction material either as building stones or to burn lime. The impact of these mining activities is enormous. Especially along the north and east coasts of the islands large parts of the coastal landscape on the Lower Terrace are completely devastated and bare of any originally occurring debris deposits. A rough estimate of how much material is removed is in the order of some million cubic meters.

Recent mining activities are taking place mainly on Bonaire, where due to the good accessibility tsunami deposits of all sizes are removed on a large scale. The boulders, which can't be moved by machines stay mostly in their original position. On Curaçao the tsunami deposits have been mined during the 19th century until the 1960's with stone crushers, which were installed along the north coast. Remnants of these stone crushers can be found for example in the area of Boka Patrick, San Pedro and east of Sint Jorisbaai. Commonly boulders of large size were blown up into smaller pieces. Tracks of transportation machines, bulldozers or smaller cars are crisscrossing the deposits (Fig. 28). After decades these tracks are showing a similar dark grayish color like the undisturbed ramparts which make it partly rather difficult to distinguish them from the natural structures. In order to protect this unique evidence for the risk of tsunami hazards in the Eastern Caribbean it is strongly recommended to establish regulations for the further removal and devastation of the tsunami deposits.

2.7.2 Economy and Infrastructure

As a result of the steep rocky shores and heavy wave and wind conditions on the windward coasts of the islands, nearly all touristic and economic infra-

Fig. 28:
Mining activities of coral and limestone debris (tsunami deposits) in Curaçao and Bonaire.

A: Spelonk area, Bonaire (aerial photograph).

B: Hato area, Curaçao (aerial photograph).

structures and coastal development areas in the Leeward Lesser Antilles are concentrated along the leeward coastlines (Figs. 29, 30). In Curaçao, the largest and most populated island with a total number of inhabitants of 130,000[2], the urbanization is concentrated in the east-central third of the island around the industrial Schottegat harbor. Offshore finance, oil industry, ship repair and maintenance, tourism, and trade (container shipment) are the mainstays of the economy (CBS, 1996). At present, about 25% of the leeward coast of Curaçao is taken up by urban, suburban and industrial development (DEBROT & SYBESMA, 2000). In the next decade this is to be expanded to about 70% (EXECUTIVE COUNCIL CURAÇAO,

1995). Bonaire's population has grown from 11,000 in 1991 to a total of 14,200 in 1995. The urbanized areas are concentrated in the central part of the island around the anchorage of Kralendijk and coastal development is limited. Bonaire's economy is primarily supported by dive tourism, and limited oil transshipment. Real-estate development and evaporative salt production are other significant economic activities. Mass tourism is the basis of Aruba's economy; about 7,500 hotel rooms stretch along the 8-kilometer coastal section from Palm Beach to Druif Bay near Oranjestad on Aruba and in total already 95% of the leeward coastline is developed. The second most important industry is the oil industry. The total population

---

[2] *Preliminary information of the Statistics (CBS) „Census 2001" indicates that the population of Curaçao dropped from 144,000 to 130,000 from 1992 to 2001.*

Fig. 29:
Oil terminal in Aruba. The refineries of the islands have been constructed in close vicinity to the coast, all with their own terminals for large tankers.

Fig. 30:
Coastal infrastructure on Bonaire. The infrastructure along extended stretches of the leeward coastlines on all islands (hotels, apartments, marinas, roads, etc.) show very high investments in a zone, which has been reached by strong tsunamis during the Holocene.

size was 83,600 by 1995 and population growth due to immigration is high. The economies of the islands have been suffering from a severe recession since 1996. Nevertheless, the islands enjoy a high per capita GNP (Curaçao: US$ 11,032 in 1995; Aruba: US$ 11,600) by regional standards. On the other hand, particularly on the islands of Curaçao and Bonaire, this income is not equally distributed among all members of the society and poverty of the rural inhabitants is an increasing problem.

3. Hurricanes and Tsunamis: General Aspects and their Relation to Coastal Forming Processes

One of the most awkward problems in evaluating the nature of a debris deposit is how to distinguish tsunami deposits from deposits as a result of hurricane-induced storm surges. In general, geomorphological and geological investigations of tsunami deposits are a relatively new research area. DAWSON (1999) stated that *"the recognition that many tsunamis deposit sediments in the coastal zone has only become an accepted idea during the last 5-10 years"*. Hitherto, no detailed investigation of the patterns and processes of sedimentation associated with a modern tsunami were carried out. As a consequence, our understanding of tsunami deposits – are they modern ones or paleotsunamis – is still relatively limited and especially paleotsunami studies are in their infancy (CHAGUÉ-GOFF & GOFF, 1999). Known occurrences of tsunami deposits fall into two main categories: fine sediment sheets and boulder beds. Whereas recent progress has been made in detailed microfossil and sediment work of fine tsunami deposits and differentiating them from paleostorm-surge deposits (DAWSON, 1999; CHAGUÉ-GOFF & GOFF, 1999, DAVIES & HASLETT, 2000), boulder accumulations attributable to tsunamis or extreme storms have been observed along a number of rocky coastlines, but a compilation of their general aspects and their relation to coastal processes is still missing. Furthermore, they can provide extremely valuable information for tsunami hazard analysis about absolute minima of runup heights or inundation distances and the magnitude of the tsunamis involved, and demonstrate tsunami energies to the public.

Therefore, the following chapters overview general aspects of tsunami and hurricanes, their genesis and characteristics as well as their geomorphologic field evidences on a worldwide to a local scale (see SCHEFFERS & KELLETAT, 2001). It is far beyond the aim of this study to describe the state of the art of hurricane and tsunami research or to discuss special questions of their geophysical background in detail.

3.1 Tsunamis – General Aspects

3.1.1 Tsunami Definition

Tsunami is a Japanese word that describes a "harbor wave". Tsunami, as a collective noun, may be singular or plural, but commonly a 's' is appended to form the plural and avoid ambiguity. The public usually thinks of a tsunami as a very large, breaking wave. The cause and nature of tsunamis is often confused through the use of the term "tidal wave", although tsunamis have no connection with tidal forcing and according to the scientific definition, it does not have to be either large or breaking. The word is now used within the scientific community for long period gravity waves generated by a sudden displacement of the water surface (DE LANGE & HEALY, 1999).

The source mechanism of the sudden displacement is normally a submarine earthquake, submarine or subaerial mass flows, large (volcanic) explosions and (sea floor) collapses and the impact of bolides in the ocean. Recently, the sudden release of gas mixtures (mostly methane) trapped within submarine sediments are suggested to be a triggering mechanism for tsunami generation as well (DFG, 2000).

This definition, used by tsunami experts in the United States, excludes meteorological-volcanic tsunamis (DE LANGE & HEALY, 1999; LOWE & DE LANGE, 2000) and storm surges, whereas these phenomena are included in the Chinese and Japanese definition and their catalogs contain both kinds of waves without any distinction made (HITTELMANN et al., 2001). A glossary of tsunami related terms are published by the UNESCO (1991).

3.1.2 Tsunami Magnitude Scales

IMAMURA-IIDA Scale

The traditional tsunami magnitude scale is the so-called IMAMURA-IIDA scale, **m**. Although the original definition was descriptive, the value is approximately equal to

$$m\,h = \log 2$$

where **h** is the maximum runup height in meters (IIDA et al., 1967). This scale is similar to the earthquake intensity scale, and is especially convenient for old tsunamis from which no instrumental records exist.

SOLOVIEV (1970) pointed out that IMAMURA-IIDA's **m** scale is more like an earthquake intensity scale rather than a magnitude. He argues, that if seismological terminology is applied to description of tsunamis, the grades of the IMAMURA-IIDA scale must be designated as the intensity of the tsunami and not its magnitude. This is because the latter value must characterize dynamically the processes in the source of the phenomenon and the first one must characterize it at some observational point. He also distinguished the maximum tsunami height and the mean tsunami height.

Another magnitude scale also based on maximum runup height, is defined and assigned for many earthquakes by ABE (1979, 1981, 1989).

Tab. 3: The IMAMURA-IIDA Tsunami Magnitude Scale.

Magnitude m	Maximum runup h [m]	Damage potential
-2	<0.3	
-1.5	0.3 - 0.5	
-1	0.5 - 0.75	Nil
-0.5	0.75 - 1.0	
0	1.0 - 1.5	Very little damage
0.5	1.5 - 2.0	
1	2 - 3	Shore and ship damage
1.5	3 - 4	
2	4 - 6	Some inland damage and loss of life
2.5	6 - 8	
3	8 - 12	Severe destruction over 400 km coast
3.5	12 - 16	
4	16 - 24	Severe destruction over 500 km coast
4.5	24 - 32	
5	>32	

Tab. 4: The tsunami intensity scale after SOLOVIEV (1978).

Intensity	Runup height [m]	Description of tsunami
I	0.5	**Very slight.** Wave so weak as to be perceptible only on tide gauge records.
II	1	**Slight.** Waves noticed by people living along the shore and familiar with the sea. On very flat shores waves generally noticed.
III	2	**Rather large.** Generally noticed. Flooding of gently sloping coasts. Light sailing vessels carried away on shore. Slight damage to light structures situated near the coast. In estuaries, reversal of river flow for some distance upstream.
IV	4	**Large.** Flooding of the shore to some depth. Light scouring on hard ground. Embankments and dykes damaged. Light structures near the coast damaged. Solid structures on the coast lightly damaged. Large sailing vessels and small ships swept inland or carried out to sea. Coasts littered with floating debris.
V	8	**Very large.** General flooding of the shore to some depth. Quays and other heavy structures near the sea damaged. Light structures destroyed. Severe scouring of cultivated land and littering of the coast with floating objects, fish and other sea animals. With the exception of large ships, all vessels carried inland or out to sea. Large bores in estuaries. Harbour works damaged. People drowned, waves accompanied by a strong roar.
?VI	16	**Disastrous.** Partial or complete destruction of man-made structures for some distance from the shore. Flooding of coasts to great depths. Large ships severely damaged. Trees uprooted or broken by the waves. Many casualties.

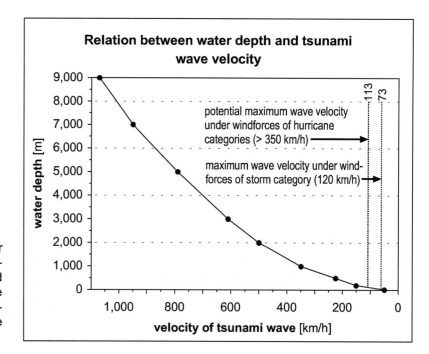

Fig. 31:
Tsunami velocity in relation to water depth. Tsunamis travel at several hundreds kilometers per hour at a speed proportional to the square root of the water depth. Consequently, they can travel across an ocean basin the size of the Pacific in less than 24 hours.

3.1.3 Tsunami Characteristics

Due to their long periods ranging from 15 to 60 minutes tsunami waves behave as shallow water gravity waves. The velocity of the propagated waves in deep water is exceptionally high and is given by:

$$C = \sqrt{gh}$$

Where **C** is the wave phase velocity, **g** is gravitational acceleration, and **h** is the water depth.

That means tsunamis travel faster in deeper water (Fig. 31). Due to their high velocities and long periods they have wavelengths of hundreds of kilometers in deep water with a maximum height of about 1.0 m, what make them difficult to recognize at the open ocean.

As the tsunami approach shallower water, the velocity decrease and the wave increase in size. The energy associated with a tsunami is distributed through the entire water column and the entire tsunami wave train. Since tsunamis are shallow water waves, the energy travels at the same velocity as the waves (DE LANGE & HEALY, 1999). As a tsunami enters shallow water it interacts with the coast in different ways:

- As a non-breaking wave that behaves like a large rapidly rising and falling tide
- As a breaking wave or bore.

This includes standing waves, the generation of edge waves, trapping or reflected waves by refraction, and the possible formation of a Mach-stem[3] along the shore (BRUUN, 1994; WIEGEL, 1976). It has long been known that the tsunami wave spectrum observed at a particular location is the result of the modification of the open ocean tsunami spectrum by the spectrum of the so-called local transfer function, which is principally a function of the topography. Not enough is known about the variation of the local transfer functions around islands, because tsunami data is scarce and episodic. These interactions are discussed in detail by CAMFIELD (1993, 1994).

In the open ocean a wind-induced wave will break when **H/L** is about 1/7. In shallow water, waves are observed to break when the water depth is 1.28 times the wave height. The distance of this depth from the shoreline depends on the steepness of the bottom. The breaking point of waves is demonstrated for the north coast of Curaçao with respect to the realistic sea bottom profile (Fig. 32).

But as periods of tsunamis may be 10-15 minutes, this can result in large runups, which have been experienced in a number of cases when a tsunami caused major damage and flooding. The theoretical runup **R** of a tsunami for smooth, impermeable slopes can be calculated according to HORIKAWA (1978):

For a slope of 1:30:

$$R = H \times 0.381(H/L)^{-0.381}$$

For a slope of 1: 60

$$R = H \times 0.206(H/L)^{-0.315}$$

---

[3] *Mach-stem: Tsunami wave traveling parallel to the coastline.*

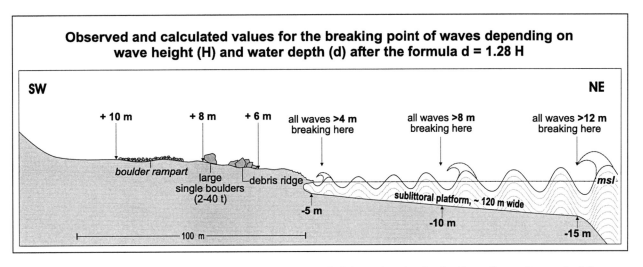

Fig. 32: The breaking point of waves is depending on wave height and water depth. Along the north coast of Curaçao all waves >12 m are breaking at the front of the subtidal platform in a distance of 120 m. At the cliff front all waves >4 m will break. Tsunami waves, however, can surge across dry beds or shores.

In comparison with field observations, the values for **R** are known to be overestimated. To overcome that problem, **H** may be replaced by **H/2** (BRUUN, 1994). This formula is not applicable at very steep slopes or cliffs, where the wave velocities are reduced abruptly and the normal forces for friction are not effective. Here, the runup height related to the wave height will be much smaller, or the wave height has been considerably higher compared to the runup value. If a tsunami hits an island, runup may be observed all around the coastline like illustrated for the island of Hawaii (Figs. 33, 34).

BRANDSMA et al. (1974) suggested with regard to shoaling and refraction effects, that tsunami behavior tend to uniformity for very long waves so that oceanic islands with a diameter smaller than the tsunami wavelength are not generally affected. Tsunami behavior becomes extremely complex when the wavelength is comparable to the size of the island (MAUL et al., 1996). Consequently, focusing and spreading of tsunami energy and amplitude occurs in response to variation in shelf depth and shoreline configuration (MURTY, 1977; MAUL et al., 1996). Where narrow shelves flank the island, the runup heights are likely to be greater than on broader shelves, where energy dissipation through shoaling over a wider approach area can occur (MAUL et al., 1996).

The most destructive tsunami are those that form a breaking bore due to the transfer of momentum to the still water trapped in front of the bore (DE LANGE & HEALY, 1999). The damage is caused mainly by the high horizontal and vertical turbulence together with the high velocity in the wave, so the available energy is more effectively to inflict damage. Flume studies indicate that a breaking tsunami generates a jet that tears off from the wave crest, followed at the point of impact by a strong splash that continues to propagate as a bore (YOUNG & BRYANT, 1992; KRIVOSHEY, 1970). This vertical turbulence of tsunami bores can lift and carry very large objects, as shown by the 1960 Chilean tsunami transporting 20 t pieces of a Japanese seawall up to 200 m inland. The current velocities generated by flooding and receding tsunami waves can be very high due to the extreme variations in water level, consequently tsunamis are very erosive. The velocity of a tsunami wave which surge onto the shore is noted by KEULEGAN (1950) and CAMFIELD (1994) as:

$$C = 2\sqrt{gh}$$

FUKUI (1963) gives a slightly lower value of

$$C = 1.83\sqrt{gh}$$

The Hokkaido Nansei-Oki tsunami produced flows with velocities of 10-18 m/s (DE LANGE & HEALY, 1999) and YOUNG and BRYANT (1992) calculated current velocities according to the equations of KIRKGÖZ (1983) of 19.9 and 18.4 m/s under flow depths of 2-10 m for the suggested tsunami event on the southeastern coast of Australia. The recent New Guinea tsunami of 1998 had an observed flow velocity of more than 70 km/h (>20 m/s) (MCSAVENEY et al., 2000). However, the interaction of reflected and refracted tsunami waves especially in regions with islands or bays can produce a very complex pattern of currents and waves, which lead to similar complex patterns of the erosional and depositional sedimentary record.

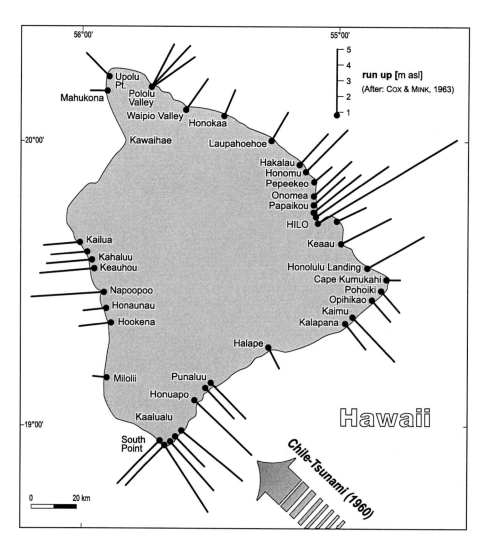

Fig. 33:
The wave of the 1960 Chile tsunami was observed all around the island of Hawaii. Note the differences and variations in wave heights over small distances (after: COX & MINK, 1963).

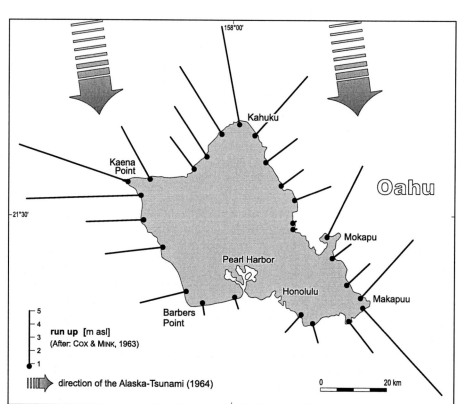

Fig. 34:
Reported runup heights for the Alaska Tsunami from 1964 on Oahu (after: COX & MINK, 1963).

3.1.4 Tsunami Generation

The generation of a tsunami wave requires the abrupt displacement of a large volume of seawater, which under the influence of gravity, returns to equilibrium. Theoretically, most of the kinetic energy of the tsunami is derived from horizontal displacements of the water column and not vertical displacement (DE LANGE & HEALY, 1999).

Earthquake tsunami generation

Shallow tectonic earthquakes generate most tsunamis, particularly subduction earthquakes are highly effective. A few tsunamis are produced indirectly by earthquake-triggered secondary mechanisms, such as mass flows or gas hydrate explosion. The minimum Richter magnitude required for a tectonic earthquake to generate a tsunami is 6.3, with a general tendency for the size of the tsunami to increase with increasing earthquake magnitude. Very large, long duration or slow earthquakes are called tsunami earthquakes, and the generated tsunami tends to be significantly larger (typically 5-10 times) than would be predicted by the magnitude of the seismic waves. Most tsunami earthquakes occur at the shallow part of subduction zones; especially slow faulting within sediments causes exceptionally large tsunamis. They may generate very stable tsunamis, which can propagate across the ocean without negligible energy loss and can cause damage at great distances from the source (teletsunami), as shown by the Chile tsunami of 1960 and the 1964 Alaska tsunami.

Landslide tsunami generation

Landslides usually are much smaller in area than earthquake-induced deformation, and often act as a point source for generating a tsunami. Landslides are well known to generate large waves and numerous examples have been documented worldwide, but only a few examples have been considered to be a tsunami formed without seismic activity.

The record of wave height associated with a landslide may be the one in Lituya Bay, Alaska on July 10, 1958 (COX & PARARAS-CARAYANNIS, 1976; LANDER, 1996). A large strike-slip earthquake (Mw = 7.9) triggered a huge rockslide starting at 110 m asl with a total volume of about 3×10^6 m³. The rockslide generated water waves that surged up the opposite slope of the fjord to 520 m altitude, resulting in the generation of a tsunami ~30 m high at the entrance of the fjord. Landslides from the flanks of the volcanoes of the Hawaiian Ridge have generated large tsunami with a runup of 375 m asl during the last interglacial high sealevel (MOORE & MOORE, 1984, 1988). Other large tsunamigenic submarine slides have been recorded in the North Atlantic Ocean, e.g. the Storegga Slide about 7200 BP (DAWSON et al., 1988). Recently evidence was found, that a breakdown of volcanic islands by slope instability may trigger – though very rare – the most destructive teletsunami in geologic history (CARRACEDO et al., 1999).

Volcanic tsunami generation

Volcanic tsunamis have small source regions, but almost a quarter of the deaths directly caused by volcanic eruptions have been attributed to associated tsunamis. This is mainly due to the distance over which the tsunami can propagate, compared to other volcanic processes. Mainly three tsunami generating mechanisms of volcanism can be named (LATTER, 1981): Mass movements (pyroclastic flows, avalanches, lahars and lava flows); cratering (submarine explosions and caldera collapse); and phase coupling (basal surges, shock waves and atmospheric pressure waves, the so-called meteorological-volcanic tsunamis). A very large volcanic tsunami (wave height >35 m), that was recorded globally, was generated by the 1883 Krakatoa eruption in Indonesia, which caused 36,000 deaths on Java and Sumatra (LATTER, 1981; SIMKIN & FISKE, 1983).

Impact tsunami generation

The tsunami generation triggered by impact of cosmic bolides is the subject of several publications (HILLS, 1998; MADER, 1996, 1998). Numerical models indicate that tsunami waves with heights of 50-100 m could be formed by bolides as small as 200 m in diameter (DE LANGE & HEALY, 1999). It is suggested that a bolide impact near Yucatan 66 million years ago produced a tsunami that was at least 90 m high along the coast of Texas and reached comparable height along the Brazilian coast.

3.1.5 Spatial Distribution of Tsunami Impacts on Different Temporal Scales

Most publications with the aim of presenting tsunami endangered coastlines or tsunami risk areas usually focus on mapping the potential tsunami sources and impact areas out of a geological or geophysical background or they produce potential inundation maps generated by numerical modeling. We want to approach an inductive presentation of the worldwide distribution of destructive tsunami impacts with illustrating only documented destructive tsunami impacts either reported by historical records and catalogues or geomorphic evidence in three different time scales. "Destructive" is defined as destruction of cliffs and barriers, transportation of boulders as well as infrastructure damage reported in historical records, like e.g. destruction of harbor fortifications and houses. The most complete worldwide tsunami catalog is archived at the National Geophysical Data Center and co-located World Data Center for Solid Earth

Geophysics (NATIONAL GEOPHYSICAL DATA CENTER, 1997). Furthermore, the location, size, runup heights and damages of past tsunamis have been compiled by several researchers and published as catalogs. For the Pacific tsunamis, IIDA et al. (1967) and LOCKRIDGE (1988) contains numerous tsunami height data as far back as 173 AD. Also Russian catalogs (SOLOVIEV & GO, 1974 a, b; SOLOVIEV et al., 1992) include Pacific tsunami data.

The records of the Mediterranean tsunami catalog reaching back to the antiquity and were compiled by PAPADOPOULOS & CHALKIS (1984). Other regional catalogs were published for Alaska (LANDER, 1996), the United States and the west coast of the U.S. (LANDER et al., 1989, 1993), Hawaii (PARARAS-CARAYANNIS, 1969), Japan (WATANABE, 1998), and Italy (CAPUTO & FAITA, 1984). ZHOU and ADAMS (1986) list historic tsunamigenic earthquake data reaching back to 1831 BC.

HITTELMANN et al. (2001) summarized interpretive pitfalls in historical hazards data sets as caused by cultural and scientific reporting changes. Most hazard catalogs, and so the tsunami catalogs, cover periods of less than 200 years and are reasonably complete and accurate for only the past 20-50 years, and consequently they are not sufficient to investigate long term hazard variations. A description of a number of different types of errors appearing in tsunami catalogs and databases is given by HITTELMANN et al. (2001), here only the pitfalls which are reflected in the tsunami distribution maps will be mentioned.

The main introduced error is associated with the lack or limited written history in different geographic regions. Whereas e.g. the first recorded tsunami in Japan occurred in 684 AD, the first recorded tsunami in the United States is dated 1788 AD. Records of tsunamis in China extend back almost 4000 years, in the Mediterranean about 2000 years. In contrast, records of tsunamis originating in the Chile-Peru coastal areas cover slightly more than 400 years, in Alaska about 200 years and those occurring in Hawaii about 180 years HITTELMANN et al., 2001).

For the Caribbean, LANDER & WHITESIDE (1997) and LANDER et al.(1998) cited the first recorded tsunami in written history for the year 1530, as expected coincident with the discovery of the American continent by Columbus in 1492. Another gap of documentation occurs comparing the records of densely populated with sparsely populated areas. Uninhabited coastlines may have experienced many unreported and, therefore, unrecorded tsunamis (HITTELMANN et al., 2001) which stress the importance of geomorphologic studies to complete the paleotsunami record.

Figure 35 shows tsunami impacts since 1900 AD, most of the tsunami are well recorded by instruments or direct observations. The time period between 1500-1899 is illustrated by Figure 36. Although the time period is extended by the factor 5, the amount of documented tsunamis is not increasing at a comparable scale due to the introduced errors in the historical tsunami catalog. This trend is accentuated in Figure 37, which overviews the time period 3000-500 BP – covering half of the Holocene with a comparable sealevel high stand as the present. The map reflects mostly those tsunami impacts, which are documented by geomorphologic evidences except for the Mediterranean and later on the East-Asian region, where long historical record exists.

3.2 Hurricanes – General Aspects

3.2.1 Hurricane Definition

A hurricane is a warm-core tropical cyclone in which the maximum sustained surface wind (using the U.S. 1-min average) is 119 km/h (74 mph or 64 kt) or more. The winds rotate counter-clockwise around a center of low atmospheric pressure (storm eye) in the Northern Hemisphere, and clockwise in the Southern Hemisphere. It derives its energy primarily from latent and sensible heat flux from the ocean, which is enhanced by strong winds and lowered surface pressure. These energy sources are tapped through condensation in convective clouds concentrated near the cyclone's center.

The terms "tropical depression" or "tropical storm" are assigned depending whether the sustained surface winds near the system's center are, respectively, ≤61 km/h and ≥62-118 km/h.

Fig. 35 *(see next page):*
Tsunami impacts since 1900 AD. The coastlines of the world, which are mostly affected by recent tsunamis are concentrated around the Pacific basin. Here, tsunamis are mainly caused by subduction earthquakes. Another area with frequent tsunami occurrence is located in the Eastern Mediterranean.

Fig. 36 *(see next page):*
Tsunami impacts covering the period from 1500-1899 AD. Extending the observed tsunami records in the past two trends become obvious: The frequency of tsunami impacts increases in most regions from single events to multiple events and areas hitherto unknown for their potential tsunami risk can be found on the map.

Chapter 3: Hurricanes and Tsunamis

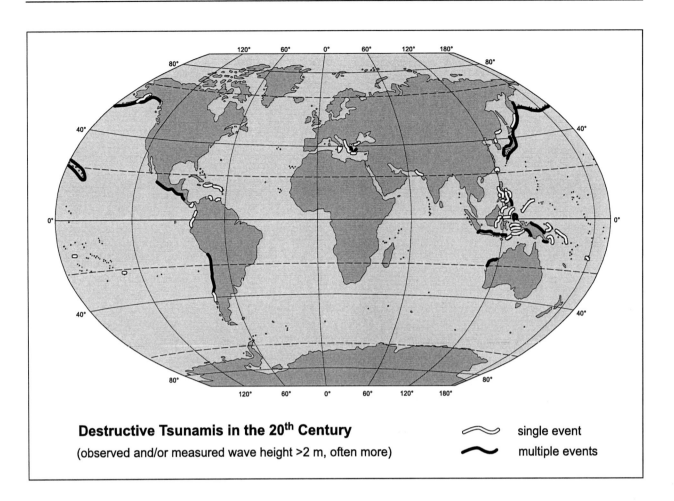

Destructive Tsunamis in the 20th Century
(observed and/or measured wave height >2 m, often more)

~ single event
— multiple events

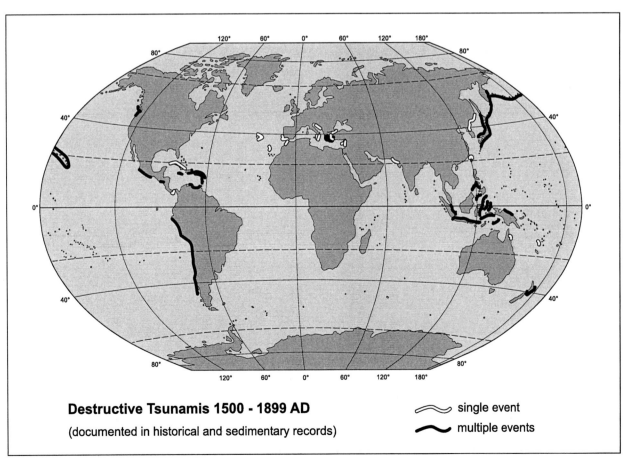

Destructive Tsunamis 1500 - 1899 AD
(documented in historical and sedimentary records)

~ single event
— multiple events

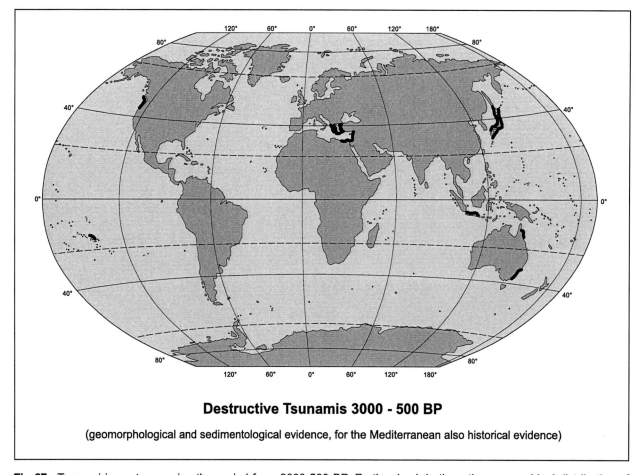

Fig. 37: Tsunami impacts covering the period from 3000-500 BP. Further back in time, the geographical distribution of observed tsunami impacts gets wider, the more sedimentary records can be deciphered.

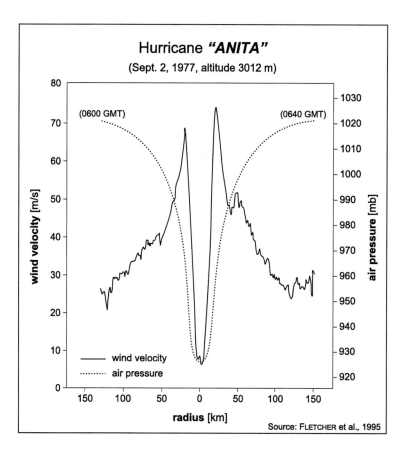

Fig. 38:
The wind and pressure profiles through hurricane *Anita* clearly show the distinct areas of the hurricane eye associated with low pressure and the ring with the highest wind velocities tangent to the eye wall (Source: FLETCHER et al., 1995).

Tab. 5: The SAFFIR-SIMPSON Hurricane Scale

Hurricane Category	Wind velocity [km/h]	Storm surge [m]
1	119 - 155	1.2 - 1.5
2	155 - 179	1.5 - 2.5
3	179 - 195	2.5 - 4.0
4	195 - 233	4 - 5.5
5	>233	>5.5

3.2.2 Hurricane Intensity Scale

The SAFFIR-SIMPSON Hurricane Scale[4] is a 1-5 rating based on the hurricane's present intensity, which gives an estimate of the potential property damage and flooding along the coast from a hurricane landfall. Sustained wind velocity is the determining factor in the scale.

3.2.3 Hurricane Characteristics

The term "tropical cyclone" refers to a cyclonic circulation, which develops over tropical or subtropical water with a temperature of at least 27°C. Frequent hurricanes occur in belts 7°-25°N and S of the equator[5]. Tropical cyclones are generally smaller in extent than extra-tropical cyclones and typically range from 200 to 1000 km in diameter at maturity (ANTHES, 1982; METEOROLOGICAL SERVICE OF THE NETHERLANDS ANTILLES AND ARUBA, 1998). The systems travel at speeds between 10 and 25 km/h and follow tracks which generally curvate away from the equator. Winds normally increase towards the center of tropical cyclones.

The sustained winds often exceed 200 km/h near the center, occasionally they exceed 300 km/h (hurricanes *Gilbert* 1988, *Hugo* 1989, *Luis* 1995) and still higher gusts may occur in well-developed systems. The winds become tangent to the eye wall at a radius of about 10-25 km (Fig. 38). The more the intensity of the circulation system increases, the narrower the band of highest wind velocities extends (Fig. 39). The speed of the anti-clockwise motion (northern hemisphere) combined with the steering speed of the storm add together on the right side of a hurricane in particular the right front quadrant, but counter each other on the left side of a hurricane (Fig. 40).

Therefore, the winds are always stronger and more destructive on the right side of a hurricane, seen in direction of the steering pass. The waves in this sector are the highest and may travel up to 1500 km in 24 hrs faster than the hurricane itself. Locations crossed by the eye experience violent winds from different directions separated by a period of calm. The wind velocity normally gradually increases during the passage of hurricane, drops down abrupt in the area of the eye and gradually decreases, when the hurricane passes the area. The gusts are highly fluctuating and the range of variations may be as much as 100 km/h (SIMPSON & RIEHL, 1981) (Fig. 41).

Within the eye, the cloud cover is low and the atmospheric pressure is less than outside of the storm. This causes a rise of sealevel near the center, which amounts to 1 cm for each millibar in atmospheric pressure (GENTRY, 1971). The mean water level can often rise up to 60 cm higher near the storm center than elsewhere. The most destructive forces of a hurricane directly at the coastline are not the wind fields, but rather this storm surge and the wind-induced waves. The maximum elevation of the storm surge is highly correlated with the zone of maximum winds in the hurricane circulation system (Fig. 42).

Storm surges associated with a severe hurricane can be as much as 3-7 m above mean water level, from which about 10% come from the low atmospheric pressure, and about 90% from the damming of wind driven water masses. The maximum water rise associated with a hurricane crossing a coast in the Caribbean has been about 5 m and occurred at Cuba (1932) and Jamaica (1722) (GENTRY, 1971). Several storms have caused storm surges of more than 3.5 m along the southern coast of the Dominican Republic. During hurricane *Camille* (1969), a SAFFIR/SIMPSON Category 5 storm and the strongest storm directly striking the United States in the twentieth century, the highest ever-measured maximum storm surge was recorded with 7.5 m (SIMPSON & RIEHL, 1981). JOHNSON (1997) calculated with a statistical approach Caribbean storm surge return periods. These storm surge values are associated with the landfall of a hurricane. In this study, the 99% storm surge limit for the 100-year value turned out to be 4066 meters. However, high water can also be caused

---

[4] http://www.nhc.noaa.gov/aboutsshs.html

[5] *The term hurricane is used for Northern Hemisphere tropical cyclones east of the International Dateline to the Greenwich Meridian.*

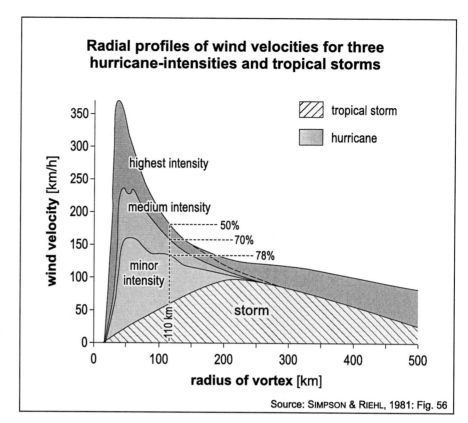

Fig. 39: The profiles through hurricanes with different intensities illustrate that the band of highest wind velocities becomes more narrow and distinct with higher hurricane intensity (Source: SIMPSON & RIEHL, 1981).

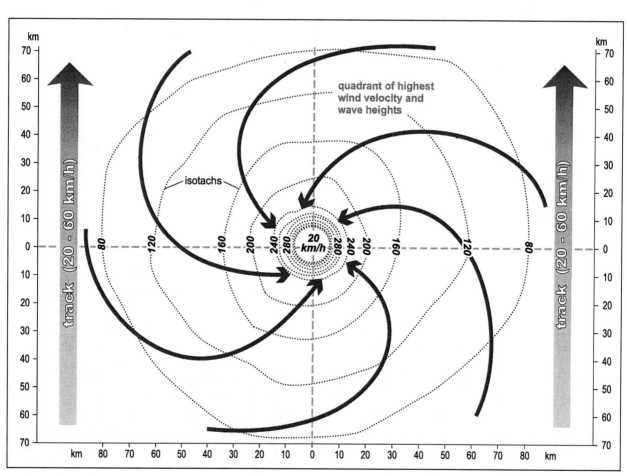

Fig. 40: The steering winds and the winds associated with the anti-clockwise motion add together on the right side of the tropical circulation system. In the upper right quadrant the strongest winds and highest waves are generated (Source: SCHEFFERS & KELLETAT, 2001).

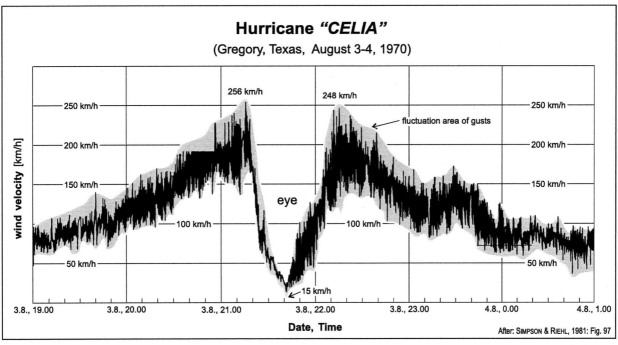

Fig. 41: Hurricane *Celia* reaches wind velocities up to 250 km/h. The bandwidth of the gusts fluctuates between 40-100 km/h. The wind velocity drops down from 256 km/h in the eye wall to 15 km/h in the center of the eye (after: SIMPSON & RIEHL, 1981).

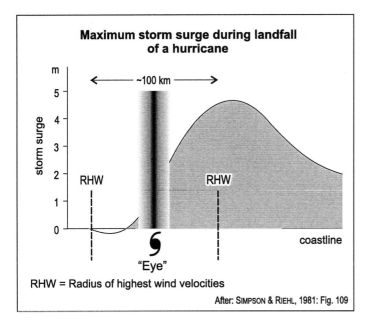

Fig. 42:
The highest storm surges are located on the right side of a tropical circulation system of the Northern Hemisphere and coincide with the highest wind velocities (after: SIMPSON & RIEHL, 1981).

by a hurricane, which moves approximately parallel to the coastline as is illustrated by hurricane *Janet* (1955) causing flooding of 2 m asl in Cuba more than 450 km removed from the center of the hurricane. Generally, the runup heights for storm waves will rarely, if ever, reach above 1.5 times the significant wave height[6] (BRUUN, 1994). On islands rising steeply from the deep ocean floor the surge is much smaller. Runups on structures and beaches are researched extensively by BRUUN (1985, 1990) and HERBRICH (1990).

It should be noted that high storm surges usually occur over a relatively short period of time (only hours) and their impact zones along the coast are also relatively narrow (100 to 150 km) (DOLAN & DAVIS, 1994). The growth of waves in a storm from the initial ripplets is governed by three factors: the wind velocity, its duration and the distance (fetch) of water surface over which it blows (DOLAN & DAVIS, 1994). With an increase in any of these factors, the height of the waves and the potential storm surge increases (Fig. 43).

---

[6] *The term "significant wave height" refers to the mean height of the highest one-third of waves measured over some time interval.*

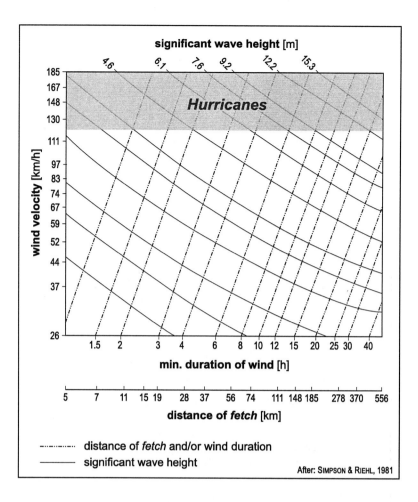

Fig. 43:
To generate a wave height of 12.2 m a wind velocity of 120 km/h (category 1 hurricane) over a period of 15 hours with a fetch of approximately 148 km is necessary. The lower the wind velocities, the longer the duration of the hurricane associated wind fields have to last to produce equivalent wave height (mod. after: SIMPSON & RIEHL, 1981).

Wind-induced waves generated by hurricanes typically range from 5-15 m, the upper limit of significant wave heights in extreme hurricanes is about 15-21 (SIMPSON & RIEHL, 1981). They reported the upper measured limit of significant wave heights offshore the coast to be 21 m during hurricane *Eloise* (1975). The highest measured open ocean waves with a height of 34 m (peak to trough) were reported by officers of the Navy oiler USS Ramapo in the North Pacific on February 7, 1933 (KLUG, 1986). Breakers of over 14 m onshore have been witnessed on the Great Barrier Reef (WOESIK et al., 1991). A phenomenon to be mentioned with extreme waves are the "freak waves", which are single waves of abnormal dimensions in height and period occurring in the midst of normal waves (BRUUN, 1994). They can be described as a coincidence of a great number of waves of unequal height and period crossing and overtaking each other, but fortunately are very rare.

These waves can reach heights of more than 25 m, but rarely occur often (GIERLOFF-EMDEN, 1980). The last reported wave height of 34 m in the North Pacific might be attributed to an extraordinary freak wave. All freak waves have a short lifetime, but their force is immense and they can cause severe local damage to coastal structures or shores.

Wind-generated hurricane waves typically have periods from 8 to 15 seconds and wavelengths from 200 to 600 m. They behave on the open ocean as deep water waves, which are found in water where the depth is greater than one half the wavelength (D>L/2). (Shallow water waves: D<L/20). The velocity **C** of a deep water wave is dependent on the wavelength and/or period:

$$C = \sqrt{\frac{g\lambda}{2\pi}}$$

where, λ = wavelength in meters, **g** = acceleration of gravity = 9.8 m/sec, π = pi (3.14), or more simply,

$$C = 1.25 \, l \, (m).$$

The greater the period or wavelength, the faster the wave velocity. Water waves have a natural tendency to travel in groups separated by regions of calm. Water waves in deep and intermediate water depths are dispersive, i.e., the waves with different wave periods travel at different speeds. According to GIERLOFF-EMDEN (1980) the highest theoretical velocity (single wave velocity) for a wave is 113 km/h (31 m/s) associated with wind velocities exceeding more than 400 km/h during extreme hurricanes. This value is significantly lower than the velocities of tsunami waves. But exactly the velocity and not the wave

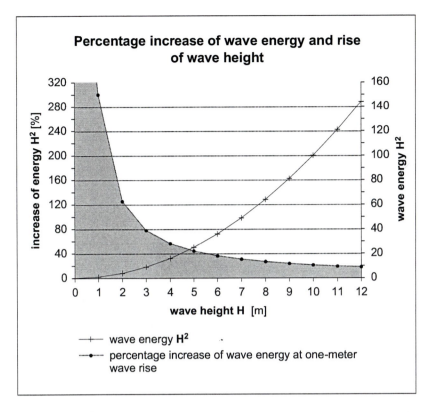

Fig. 44:
Relation between the wave height and the energy of a wave in 1 m steps. The increase in wave heights with regard to the geomorphic effects plays a more important role in lower wave heights categories, whereas an increase of wave height in higher wave classes is of less geomorphic relevance.

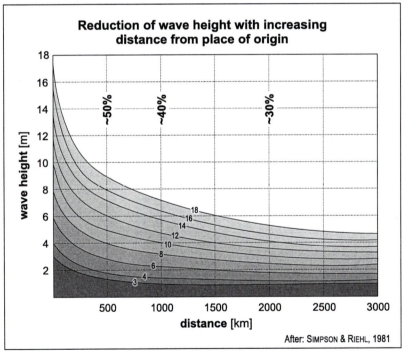

Fig. 45:
Relation between wave heights and the distance to the area of wave origin and the origination of oceanic swell (after: SIMPSON & RIEHL, 1981).

energy, as might be assumed, is the critical variable in determining if a wave is capable of overturning and moving a given object and in so far of more of geomorphic relevance. Consequently, storm waves are limited in their transporting capacity due to their considerably lower velocity compared to tsunami waves.

The total energy of a storm wave is proportional to the square of the wave height. Doubling **H** increases the wave energy by a factor of 4.

- E is proportional to H^2
- where **E** = total energy of a wave, and **H** = wave height

It can be seen that the relative increase of wave energy is of more geomorphic significance if wave heights are lower, e.g. the energy of a wave of 10 m height is 100 times higher than the energy of a 1 m wave, whereas a doubling of the wave height to 20 m will only increase the wave's energy by the factor 4 (Fig. 44).

Tab. 6: The main characteristics of hurricane and tsunami waves as reported by field observations.

	Hurricanes	Tsunamis
frequency	high	low
duration	hours to days	minutes to hours
rise in water level	slow to moderate	fast
energy	moderate	extreme high
impact	local	local to distant
Open ocean:		
wave height [m]	12 - 25	0.5 - 1.0
wave length [L]	200 - 600 m	100 - 300 km
wave period	8 - 15 sec	Minutes to hours
wave velocity [km/h]	40 - 60 (max. 113)	500 - 1,000
Shore zone:		
max. wave height [m]	6 - 9	>30 m
max. storm surge/run up [m]	7 - 7.5	>100 m
wave velocity [km/h]	20 - 30	100 - 200

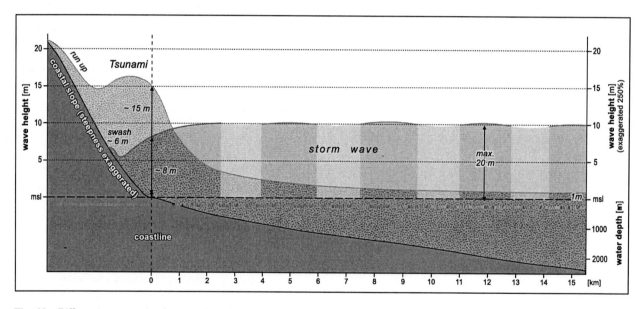

Fig. 46: Different approach of tsunami and storm waves at a cliff coast.

With increasing distance to the area of their origin, both – wave height and energy – are decreasing significant, in particular within the first hundreds of kilometers (Fig. 45).

Summarizing the above, Table 6 overviews the main characteristics of hurricane and tsunami waves. Figure 46 illustrates the different approach of tsunamis and storm waves at a cliff coast.

3.3 Hurricanes or Tsunamis – Potential Capacity and Field Observations of Debris Deposition

Generally there are several reasons for the occurrence of debris material (pebbles to boulders) in coastal environments:

- Terrestrial chemical deep weathering associated with in situ forming of boulders ("woolsack-weathering"), coming into the surf environment by sealevel rise or land subsidence. Examples can be found in SW- and SE-Australia, Tasmania, around Rio de Janeiro in Brazil, or on the Seychelles Islands. On Aruba there are some examples around the northernmost point near "California" as well as at the East Coast north and south of "Natural Bridge", but beside these examples in restricted areas, the described process can be excluded to be a source of the described debris material.

- Transportation of debris material into the surf zone from the terrestrial environment occurs either by falling or sliding from steep slopes and cliffs caused by gravity, or by transportation through rivers and creeks, as in many climates with mechanical weathering. Examples of this mechanism can be found a the beaches of small bokas on all ABC-islands, where basalt or diabas boulders have been transported by strong run off along short and steep creeks. But generally this process can be neglected as a source of the debris material due to the fact that the next cliff edge in most cases is located several hundred meters far more inland than the boulder ramparts and ridges.

- In cold climates coarse material usually is widespread produced by frost weathering or glacial deposition. Evidently these processes have not occurred on the ABC-islands.

- Debris material may be produced in the coastal zone itself, in particular in foreshore areas with coral reefs affected either by bioerosion or hazards like storms or tsunamis. Another source area is the cliff front or the supratidal zone, which might be destructed by wave impact. This process can produce debris of significant size during tsunami events.

Effects that are more dramatic are associated usually with a more severe impact. However, there are some exceptions: For example, a hurricane may be strong, but if there is no source of debris due to the activity of an earlier hurricane, than no storm deposits can be built up and the geomorphologic evidence will be relatively less (e.g. *Gilbert* 1988 after *Allen* 1980 on Jamaica (WOODLEY, 1990); and *Betsy* 1965 after *Donna* 1960 on Florida Keys (BALL et al., 1967; PERKINS & ENOS, 1968). Furthermore, the availability of coral debris is controlled by the size (age) of the corals, which in turn is a function of the time elapsed since the last devastating impact and the rate of recovery of the species in that specific environment (KJERFVE et al., 1986; HARMELIN-VIVIEN, 1994).

3.3.1 Hydrodynamic Aspects of Boulder Movement

Research concerning the carrying capability of waves, especially with regard to the application of coastal protection, is classically the field of hydrodynamic studies. It is far beyond the scope of this study to discuss the hydrodynamic theories of wave transport, the interested reader is referred to the standard publications by NEWMAN (1977) and SARPKAYA & ISAACSON (1981). A more detailed overview of the dynamics of boulder movement is presented in SCHEFFERS & KELLETAT (2001), nevertheless the most important literature concerning the subject should be shortly presented.

LORANG (2000) tries to simulate and model the realistic conditions for boulder movement via 46 theoretical steps, where e.g. parameters like block weight, shape, roughness, beach slope, wave height and -period, swash velocity or the packing arrangement are integrated. The forces and parameters, determining the movement of boulders, are simplified in Figure 47.

Theoretically, the following relations between the wave height of unbroken waves and the actual movement of round-shaped boulders were predicted (Table 7).

It has to be noticed that only the horizontal movement of boulders was subject of the study without considering vertical dislocations. An important factor affecting the transportation capability of waves is the shape of the boulder. As Figure 48 shows, the volume and weight of a boulder increases rapidly with only a slight increase in diameter, in particular if the boulder is of more cubic shape. The amount of energy, which is necessary to move angular shaped boulders versus spherical shaped ones, is described proportional as 5:1.

The only empirical approach was carried out by OAK (1985) on a boulder beach in Australia during a 2-year field survey, where the movement of boulders was observed while 11 storms passed the area. The results are presented in Table 8. From these results it can be deduced, that the relationship between wave height in the open sea and the weight of transported boulders is not linear, the experiment rather shows that if a wave exceeds a certain threshold in height (here: >4.2 m) unexpectedly boulders of higher weight are moved. This important observation clearly is inconsistent with the theoretical predictions.

An interesting study by MASSEL & DONE (1993) focuses on the effects of cyclone waves on massive

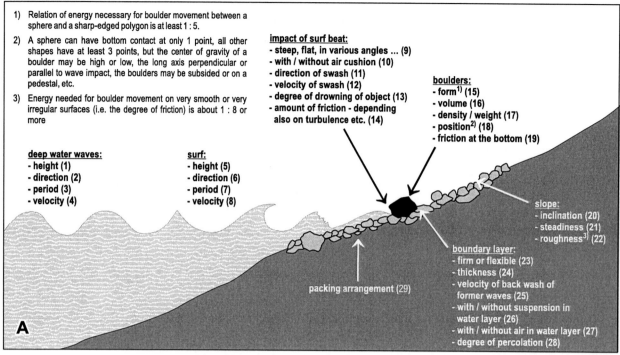
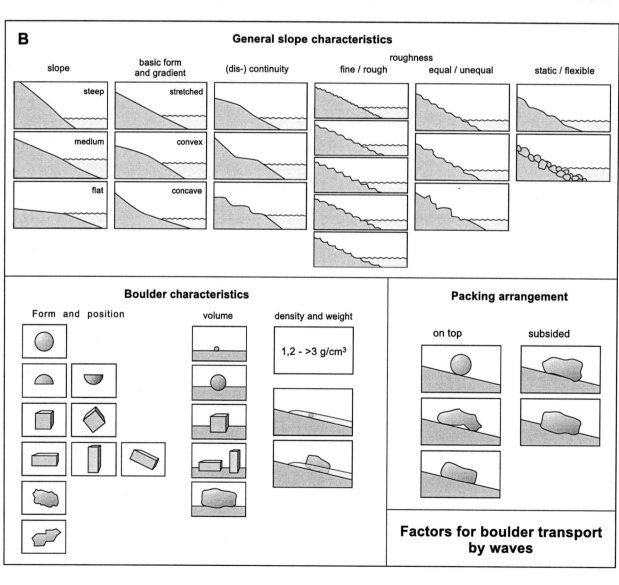

Factors for boulder transport by waves

Wave height [m]	Boulder weight [t]
1	0.25
2	0.3
3	0.4
6	0.5 - 0.7
8	0.55 - 1.1

Tab. 7:
Relation between wave height and the weight of boulders moved by this particular wave height (LORANG, 2000).

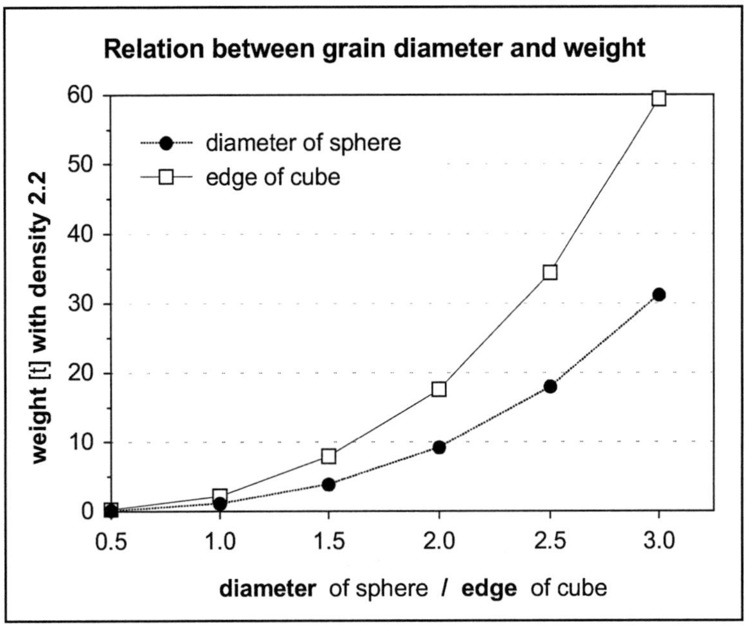

Fig. 48:
Relation between the weight of a boulder and its shape – round or cubic. The weight of a cubic boulder increases rapidly with increased diameter.

coral assemblages on the Great Barrier Reef. The analysis of shear, compression and tension forces generated by waves indicate that corals firmly attached to solid substratum, even if only over a small portion of their base, can resist all waves, regardless of colony size or shape, cyclone intensity or region. They stated, that corals on the GBR which are dislodged in storms must, therefore, be either unattached or weakened by borings, or strongly attached to substrata which are themselves weak, friable, or of small mass. The threshold wave height needed to overturn a coral of a given diameter is found to be associated with the specific return period of cyclonic conditions, which induce waves of a specific height. The results can be summarized as follows: At 3 m depth only attached coral can persist. For sites at a depth of 12 m, the threshold point corresponds to corals of 40 cm in diameter, irrespective of local variations. However, in 9 m depth it varies between 40-80 cm diameter depending among the region. At 6 m depth, the point of deviation corresponds to corals of 60 cm diameter. The authors pinpoint that a higher wave height (larger force) is needed to overturn a coral at a greater depth, as would be expected.

NOTT (1997) adapted the formulas used by MASSEL & DONE (1993) with incorporating the lift force and calculated the critical wave height necessary to overturn boulders according to the following equations for tropical cyclone generated waves (a) and tsunami waves (b):

(a) $H = (p_s - p_w / p_w) 8a / [(ac/b^2)C_d + C_l]$

(b) $H = (p_s - p_w / 4p_w) 2a / [(ac/b^2)C_d + C_l]$,

where p_w = density of water at 1.02 g/ml, p_s = density of boulder at 2.5 g/cm³, C_d = coefficient of drag = 1.2, C_l = coefficient of lift = 0.178, **a** = A-axis of boulder (long axis), **b** = B-axis of boulder (intermediate axis), **c** = C-Axis of boulder (short axis).

The accuracy of the wave height equations was tested by comparing calculated and observed wave heights and the size of transported concrete slabs by the June, 17th 1998 tsunami event in Sissano, Papua New Guinea and during the passage of tropical cyclone *Vance* in Western Australia on March 22nd 1999 (NOTT, 2000).

Fig. 47 *(see previous page):*
The factors, which influence the movement of boulders in nature are extremely various.
A: 29 parameters illustrate exemplary the complexity of the boulder movement process.
B: The bandwidth of possible variations for some factors.

Tab. 8: Relation between wave height, wave period and the movement of boulders (OAK, 1985).

Wave height offshore [m]	Wave periods offshore [sec]	Largest boulder moved [kg]
2.1	10	28
2.2	12	53
2.2	11	93
2.3	11	24
3.2	12	27
3.3	11	39
4.2	14	24
4.3	14	2191
5.1	15	898
5.4	11	1276
6.0	12	935

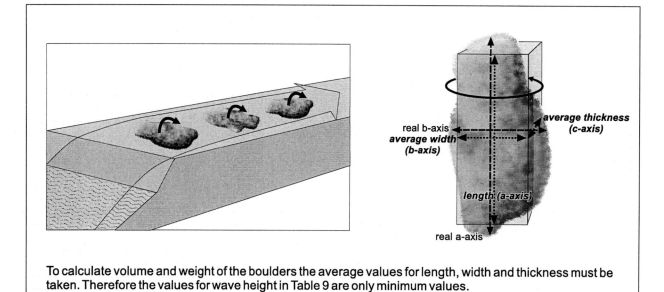

To calculate volume and weight of the boulders the average values for length, width and thickness must be taken. Therefore the values for wave height in Table 9 are only minimum values.

Fig. 49: Illustration of boulder movement and measuring method to calculate boulder volume and weight.

Table 9 lists the wave height required to overturn boulders according to these equations for the measured boulders of this study. Rockpool features now located on the basis or the sides of the blocks can proof evidence of overturning besides simple sliding during the transport of the boulders. For the calculation of volume and weight we used the averaged values for length, width and thickness. The real dimensions of the axes exceed the averaged values, and insofar the wave heights calculated in Table 9 can considered to be minimum values.

The results demonstrate that the possibility remains that extremely large hurricane waves may have the capability to overturn boulders of an insignificant quantity, but considering their present position in cases up to 12 m asl, it seems to be unlikely that such waves would be able to deposit them into their present position. From the measured 76 distinctive boulders on Curaçao (weight >1 t) – except very few – all require storm wave heights which never have been observed at any coastline of the world (up to 125 m!). For the 42 measured boulders on Bonaire none could be moved by storm surf regarding the required waves height of 14-89 m high. Even on Aruba, where the boulders usually are much smaller, waves of 13-56 m would be needed. In contrast, the wave height calculated for tsunamis are well in the range of observed events.

The table shows that waves of lesser wave heights were capable of transporting boulders of greater volume. This is because the critical variable for the initial transport of boulders is the length of the C-axis, which is inversely proportional to the required drag force. The smaller the area of the C-axis exposed,

Chapter 3: Hurricanes and Tsunamis

Tab. 9: Dimensions of selected boulders on the ABC-islands and height of waves (tropical cyclone and tsunami) required to overturn these boulders. For calculation of wave heights the equations of NOTT (1997) were applied.

	a-axis	b-axis	c-axis	volume [m³]	weight [t]	wave height [m]	
						hurricane	tsunami
Curaçao	200	50	40	0.4	1.0	6	2.2
	100	100	100	1.0	2.5	9	3.3
	100	100	100	1.0	2.5	9	3.3
	100	100	100	1.0	2.5	9	3.3
	120	100	100	1.2	3.0	9	3.3
	530	100	100	5.3	13.3	10	3.6
	260	60	30	0.5	1.2	11	4.2
	190	110	100	2.1	5.2	11	4.1
	140	110	90	1.4	3.5	12	4.4
	160	110	90	1.6	4.0	12	4.5
	240	110	90	2.4	5.9	12	4.6
	290	120	100	3.5	8.7	13	5.0
	150	120	100	1.8	4.5	13	4.7
	210	120	90	2.3	5.7	14	5.4
	220	120	90	2.4	5.9	14	5.4
	500	120	100	6.0	15.0	14	5.2
	190	130	100	2.5	6.2	15	5.6
	260	150	140	5.5	13.7	15	5.5
	310	150	130	6.0	15.1	16	6.0
	260	110	70	2.0	5.0	16	5.9
	180	130	90	2.1	5.3	16	6.1
	130	110	60	0.9	2.1	16	6.1
	130	130	80	1.4	3.4	17	6.4
	210	130	90	2.5	6.1	17	6.2
	300	150	120	5.4	13.5	17	6.4
	230	180	170	7.0	17.6	17	6.3
	200	170	150	5.1	12.8	17	6.3
	250	180	160	7.2	18.0	18	6.7
	400	200	200	16.0	40.0	19	7.0
	310	160	120	6.0	14.9	19	7.2
	130	100	40	0.5	1.3	19	7.3
	180	150	100	2.7	6.8	19	7.1
	300	150	110	5.0	12.4	19	6.9
	200	190	140	5.3	13.3	22	8.1
	330	140	80	3.7	9.2	22	8.2
	400	170	120	8.2	20.4	22	8.3
	200	130	60	1.6	3.9	23	8.7
	260	160	90	3.7	9.4	24	9.1
	370	200	150	11.1	27.8	24	9.0
	330	190	130	8.2	20.4	25	9.2
	690	200	150	20.7	51.8	25	9.4
	210	150	70	2.2	5.5	26	9.8
	250	110	40	1.1	2.8	26	9.6
	190	150	70	2.0	5.0	26	9.6
	210	200	120	5.0	12.6	27	10.1
	450	230	170	17.6	44.0	28	10.5
	300	200	120	7.2	18.0	29	10.7
	150	130	40	0.8	2.0	30	11.1
	340	170	80	4.6	11.6	31	11.6
	320	160	70	3.6	9.0	31	11.7
	340	220	130	9.7	24.3	32	12.0
	240	130	40	1.2	3.1	34	12.5
	200	190	80	3.0	7.6	34	12.6
	350	200	100	7.0	17.5	34	12.8
	200	170	60	2.0	5.1	35	13.2
	400	180	80	5.8	14.4	35	13.1
	540	280	200	30.2	75.6	35	13.2
	150	130	30	0.6	1.5	36	13.5
	250	180	70	3.2	7.9	36	13.6
	800	250	150	30.0	75.0	39	14.4
	350	220	100	7.7	19.3	40	15.0
	360	260	140	13.1	32.8	40	15.0
	400	290	180	20.9	52.2	40	14.9
	200	180	50	1.8	4.5	44	16.3
	500	300	180	27.0	67.5	44	16.3
	230	160	40	1.5	3.7	45	16.9

55

Tab. 9: continuation

	a-axis	b-axis	c-axis	volume [m³]	weight [t]	wave height [m] hurricane	wave height [m] tsunami
	280	190	60	3.2	8.0	46	17.0
	300	210	70	4.4	11.0	48	17.9
	530	290	150	23.1	57.6	48	18.1
	300	250	100	7.5	18.8	48	17.8
	280	280	120	9.4	23.5	49	18.1
	210	200	50	2.1	5.3	51	19.1
	290	270	100	7.8	19.6	53	19.8
	350	220	70	5.4	13.5	53	20.0
	500	400	250	50.0	125.0	54	20.1
	470	340	170	27.2	67.9	56	20.9
	590	530	360	112.6	281.4	65	24.4
	700	390	200	54.6	136.5	65	24.5
	340	340	120	13.9	34.7	68	25.3
	650	460	260	77.7	194.4	69	25.6
	370	320	90	10.7	26.6	78	29.2
	550	350	120	23.1	57.8	80	29.9
	500	420	160	33.6	84.0	83	31.0
	380	320	70	8.5	21.3	93	34.8
	710	350	100	24.9	62.1	98	36.4
	480	350	60	10.1	25.2	125	46.7
Bonaire	300	200	270	16.2	40.5	14	5.2
	420	200	200	16.8	42.0	19	7.0
	250	220	220	12.1	30.3	19	7.3
	560	190	160	17.0	42.6	21	7.9
	420	200	180	15.1	37.8	21	7.7
	250	250	250	15.6	39.1	22	8.1
	210	190	130	5.2	13.0	23	8.7
	290	220	190	12.1	30.3	23	8.4
	550	250	240	33.0	82.5	24	9.1
	550	250	220	30.3	75.6	26	9.9
	560	260	240	34.9	87.4	26	9.8
	380	260	200	19.8	49.4	30	11.1
	400	220	130	11.4	28.6	33	12.2
	500	270	190	25.7	64.1	34	12.9
	310	200	100	6.2	15.5	34	12.5
	420	250	150	15.8	39.4	36	13.6
	550	260	170	24.3	60.8	36	13.4
	320	260	150	12.5	31.2	37	13.9
	420	350	280	41.2	102.9	38	14.1
	600	340	270	55.1	137.7	39	14.5
	400	370	300	44.4	111.0	39	14.6
	480	320	220	33.8	84.5	41	15.2
	530	300	190	30.2	75.5	42	15.6
	520	260	140	18.9	47.3	42	15.8
	400	250	120	12.0	30.0	44	16.3
	400	320	180	23.0	57.6	47	17.5
	360	240	100	8.6	21.6	47	17.4
	530	330	200	35.0	87.5	47	17.6
	570	320	190	34.7	86.6	47	17.6
	550	260	120	17.2	42.9	49	18.3
	560	320	180	32.3	80.6	49	18.5
	800	350	200	56.0	140.0	55	20.5
	600	370	190	42.2	105.5	61	22.8
	500	410	220	45.1	112.8	62	23.3
	950	350	180	59.9	149.6	62	23.0
	330	330	120	13.1	32.7	64	24.1
	550	450	260	64.4	160.9	64	24.0
	900	350	160	50.4	126.0	68	25.4
	530	490	220	57.1	142.8	84	31.2
	500	480	200	48.0	120.0	86	32.1
	600	450	180	48.6	121.5	88	32.9
	850	410	160	55.8	139.4	89	33.1
Aruba	300	130	120	4.7	11.7	13	4.9
	260	160	120	5.0	12.5	19	7.1
	320	160	120	6.1	15.4	19	7.2
	280	160	80	3.6	9.0	27	10.2
	470	140	60	3.9	9.9	30	11.1
	200	160	50	1.6	4.0	37	13.9
	410	150	50	3.1	7.7	39	14.4
	300	180	50	2.7	6.8	49	18.3
	270	180	40	1.9	4.9	56	20.9

the greater is the required drag force assuming that boulders are overturned through the B-Axis or the A-axis parallel to the wave front (NOTT, 2000).

Still the possibility remains that extremely large hurricane waves may have the ability to overturn boulders of smaller size, but it seems to be unlikely that such waves are capable to deposit them into their present position sometimes up to 12 m asl. Therefore, a tsunami would appear to be the most capable and plausible wave to transport boulders of the size observed along the coastline of the Netherlands Antilles and Aruba.

3.3.2 Hurricanes and Tsunamis – worldwide

One of the most commonly reported features of coastal debris deposition is the storm ridge, or rubble rampart of coral shingle that accumulates near the seaward margin of the coastline, especially in areas where coral reefs are situated. Usually they were attributed to a severe hurricane event and only recently the deposits may be discussed as a consequence of possible tsunami impacts. SCOFFIN (1993) overviews the geological and geomorphological effects of hurricanes on coral reefs and their interpretation as storm deposits. The deposition of rubble ramparts after a storm was reported by numerous authors (HARMELIN-VIVIEN, 1994): STEPHENSON et al. (1958) and STODDART et al. (1978) at Low Isles, GBR; MCKEE (1959) and BLUMENSTOCK (1961) at Jaluit atoll, Marshall Islands; STODDART (1963, 1971) in British-Honduras; OGG & KOSLOW (1978) in Guam; KOHN (1980) at Eniwetok atoll, Micronesia. Several coral rubble ridges parallel to the outer rim of islets were observed on different French Polynesian atolls, testifying by their difference in color and height of different cyclonic events (HARMELIN-VIVIEN, 1985; PIRAZOLLI et al., 1988). MARAGOS et al. (1973) and BAINES et al. (1974) described a boulder rampart, 3-4 m high, 19 km long and about 37 m wide on Funafuti atoll after cyclone *Bébé* (1972). They calculated an estimated volume of 1.4×10^6 cubic meters and a mass of 2.8×10^6 tons. BAINES & MCLEAN (1976) and MCLEAN (1992) demonstrated that Funafuti atoll has grown episodically during the last 2000 years and coral rubble deposits form one third of the island.

The morphology of storm ridges is controlled in part by the orientation and nature of the waves, which are influenced by the morphology of the reef as well as by the size and shape of the accumulated debris (SCOFFIN, 1993). The rubble ridges in the Pacific region are usually up to 2-3 m high and have a width of 5-7 m, in Jaluit and Jabor they are located 5-10 m offshore (STODDART, 1971). The storm ridges typical of the low wooded island type of platform reef on the inner shelf of the Great Barrier Reef may extend right round the platform reef, but usually occur just on the windward side, where they are 20-70 m wide and 1-4 m high (SCOFFIN, 1993). They are normally rich in platy and rod-shaped coral fragments with some rare boulders present. The platy and rod-shaped fragments show an alignment normal to the wave front direction and an imbrication of stacked fragments dipping to seaward. On the leeward flanks of some inner shelf platform reefs, the ridges consist essentially of fragments of massive coral colonies 10 cm to 2 m in diameter. On some reefs, two discrete boulder tracts are developed, which did not move since they were deposited as indicated by the position and height of intertidal encrusting and endolithic organism (SCOFFIN, 1993).

Exceptionally large boulders of coral or reef framework are found near the windward margin of some reefs with around 50 m³ in volume, which he suggested to be left as lag deposits accompanied by earlier storm ridge deposits of shingle, which was subsequently transported away (SCOFFIN, 1993). BAINES & MCLEAN (1976) used painted boulders as tracers to detect movement of debris under normal swell with wave height of 1.5 m on the *Bébé* ridge in Funafuti. Debris of 20-30 cm was moved tens of meters along and over the ridge, whereas debris in excess of 1 m never was moved. Two other responses of massive corals have been observed in cyclone-damaged areas on the GBR (DONE & DEVANTIER, 1986; WOESIK et al., 1991; DONE, 1992) and on Vanuatu, SW Pacific Ocean (DONE & NAVIN, 1990).

The dislodgment of corals (*Porites* sp.) up to several meters in diameter and height (no exact measurements), but they either remained at or near their position of growth or they came at rest down-slope. The capability of storm surge and storm waves to transport boulders of various sizes was observed by NOTT (2000) when tropical cyclone *Vance* crossed Western Australia in March 1999. Historically for this region, *Vance* was the most intense tropical cyclone with a surge between 2.5-4 m and wind velocities of 300 km/h near the eye. NOTT (2000) reported that the largest boulders moved where considerably smaller than 0.6 m³. FLETCHER et al. (1995) described a boulder of approx. 300 kg, which was deposited up to a height of +6 m during hurricane *Iniki*. The largest storm boulders recorded are those at the Rangiroa atoll, where highly cemented blocks of reef rock up to 6 m in height are jumbled on the reef flat (STODDART, 1971). These are thought to date from a hurricane in about 1900, when according to the author the hurricane conditions must have greatly exceeded those of any recent hurricane. Two other examples for boulder movements by storms outside the hurricane belt should be mentioned: BASCOM (1959) reported

Region	Origin	Horizontal displacement [m]	Vertical displacement [m]	Altitude [m asl]	Boulder weight resp. -size	Equals cube with edge length of	REFERENCE
1. Oregon (USA)	Storm	?	>30	–	135 pounds	0.3 m	BASCOM, 1959
2. South Carolina (USA)	Hurricane	15	3	3	180 kg	0.42 m	STAUBLE et al., 1991
3. Kauai (Hawaii)	Hurricane	50	?	6	~300 kg	0.5 m	FLETCHER et al., 1995
4. Puerto Rico (USA)	Hurricane	20	3	2.5	1.2 t	0.75 m	BUSH, 1991
5. Portugal	Tsunami	?	?	3	6 - 8 t	1.4 m	DAWSON, 1996
6. New South Wales (Australia)	Tsunami	?	33	33	6 m³	1.8 m	BRYANT et al., 1996
7. Cyprus	Tsunami	80	10	8	28 t	2.25 m	KELLETAT & SCHELLMANN, 2000
8. Curaçao (Netherlands Antilles)	Tsunami	50	7	8	30 t	2.3 m	KELLETAT & SCHEFFERS, 2001
9. Grand Cayman (Caribbean)	Tsunami ?	100	8	10	40 t	2.5 m	JONES & HUNTER, 1992
10. Curaçao (Netherlands Antilles)	Tsunami	8	3	2	70 t	3.05 m	KELLETAT & SCHEFFERS, 2001
11. New South Wales (Australia)	Tsunami	?	?	40 - 50	90 t	3.3 m	BRYANT et al., 1996
12. Ryukyu (Japan)	Tsunami	?	?	10	100 m³	4.6 m	DAWSON, 1994
13. Queensland (Australia)	Tsunami	70	0 - 2	0 - 2	up to 290 t	4.7 m	NOTT, 1997
14. French-Polynesia	Tsunami	~70	~2	2	1,000 m³	10 m	TALANDIER & BOURROUILH-LE-JAN, 1988

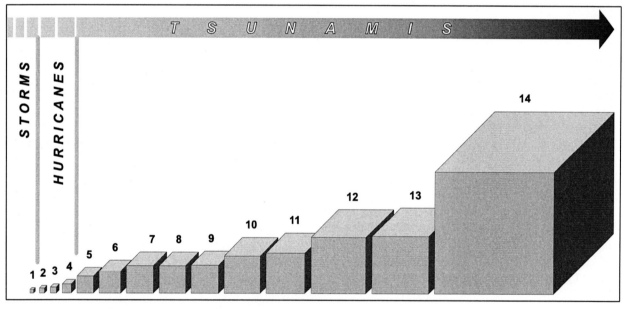

Fig. 50: Exemplary field evidences of boulder deposition by hurricanes or tsunamis.

that during a severe storm along the coast in Oregon (USA) a particle of 65 kg was thrown 30 m high into a lighthouse (see Fig. 50). The sideward movement of 6 t-tetrapods during a heavy storm at Sylt (Germany) was observed by KELLETAT (1989). Summarizing it can be stated, that in no case of observed hurricane impact large reef boulders of cemented Pleistocene limestone material have been broken out of a cliff front and transported onto a reef flat. Only in the past 15 years awareness has been heightened that most plausible tsunami waves were responsible for the deposition of debris accumulations of large boulders or extensive rubble ridges. One of the first studies discussing a possible tsunami impact was carried out by BOURROUILH-LE JAN & TALANDIER (1985), who observed on Rangiroa in the Tuamoto archipelago areas of coral reefs, where large numbers of giant boulders up to 750 m³ appear to have been transported across the atoll rim and into the lagoon. They considered paleotsunamis as likely depositional agents, but it is also speculated that storm surges associated with former hurricanes may have been responsible for the boulder deposition (TEISSIER, 1969; PIRAZZOLI et al., 1988). Perhaps the most famous example are the coral gravel and boulder beds from the Hawaiian islands first described by MOORE & MOORE (1984, 1988) and MOORE et al. (1992).

They are interpreted as the products of giant tsunamis with wave runups of up to 375 m, caused by a large submarine slide/debris avalanche (the Alika phase 2 event) on the western flank of the Mauna Loa volcano circa 105,000 years ago. These deposits of about 8 m thickness, known as the Hulopoe Gravel, are discussed controversial, not least because of the alternative suggestion that they are remnants of beaches displaced by uplift movements of the island during their deposition. Most boulders with diameters of approx. 0.5 m occur no higher than 50 m asl (MOORE & MOORE, 1984).

Another Pleistocene tsunami impact on the island of Molokai, Hawaii was dated approx. 200,000-240,000 BP by MOORE et al. (1994), where boulders of 10 m^3 were transported several kilometer inland. Recently in Hawaii, during the tsunami of 1960, a boulder of 22 t was transported 200 m inland (DOMINEY-HOWES, 1996). KELLETAT & SCHELLMANN (2001) reporting tsunami deposits along the East coast of Big Island Hawaii and Oahu with weights up to several tons and located up to 10 m high and 200 m inland. KELLETAT (pers. comm., 2000) reports not published preliminary observations on tsunami boulder dislocations on the barrier reefs of the Palau archipelago (Micronesia), in SW-Australia around Albany, in NW-Australia along the Ningaloo Reef near Exmouth, in Victoria (Australia), or on Long Island, Bahamas.

Boulder fields have been observed in the southern Ryukyu Islands, Japan (OTA et al., 1985; KAWANA & PIRAZZOLI, 1990; NAKATA & KAWANA, 1993). Here, numerous boulders up to 100 m^3 occur up to 30 m asl and have been interpreted as tsunami deposits during the Holocene. The recent Hokkaido-tsunami from 1993 transported several small boulders (DAWSON, 1996). Several tsunami impacts have been described for the southeastern coast of Australia. Numerous giant boulders are reported to have been removed from pre-existing emerged coastal rock platforms by YOUNG & BRYANT (1992). BRYANT et al. (1996) describes boulders of up to 50 m^3 with a weight of 90 t deposited along the same coast and suggested that tsunamis act as a major control on coastal evolution.

NOTT (1997) reports the largest tsunami boulders with a weight of up to 290 t on the Australian continent along the coast of Cairns inside the Great Barrier Reef.

From the Mediterranean tsunami boulder deposition is documented by HECK (1947) where in the Strait of Sicily during the tsunami of 1908 a wave of 11-13 m transported a boulder of 20 t approx. 20 m further sideward. A boulder of 6-8 t was moved up to 3 m by the Lisbon-tsunami (EDWARDS, 1842 in: DAWSON, 1996). MASTRUNOZZI & SANSO (2000) are describing large boulders up to 80 t, which were transported by a tsunami approx. 40 m inland and to a height of 1.8 m presumably between 1421 and 1568 AD at the coastline of Apulia/Italy. The most detailed field survey of tsunami impact in the Mediterranean was carried out by KELLETAT & SCHELLMANN (2001a), who found evidence for a severe tsunami impact on Cyprus dating back only 200-300 years, with movement of thousands of large boulders onshore, some of them exceeding 20 tons.

3.3.3 Hurricanes and Tsunamis in the Intra Americas Sea (IAS[7])

Hurricane record

The official Atlantic hurricane season extends from June, 1st to November, 30th with the large majority of intense hurricanes occur in just the three month of August, September and October (LANDSEA, 1993). Over a 110-year period covering 1886 to 1995, a total of 956 tropical storms (of which 543 reached hurricane force) have been recorded over the North Atlantic Area (Fig. 51) (JARVINEN et al., 1984; METEOROLOGICAL SERVICE OF THE NETHERLANDS ANTILLES AND ARUBA, 1998).

While these records are available for the entire Atlantic Basin for hurricanes back to the late 1800s and for landfalling hurricanes back to the 16th century (LUDLUM, 1989), reliably knowledge about the intensity of such systems extends for a much briefer period of time. For the whole Atlantic Basin, reliable intensity measurements exist back only to the beginning of routine aircraft reconnaissance in 1944, but even these data have been corrected subsequently (LANDSEA, 1993; NEUMANN et al., 1993, 1997). READING (1989) and GRAY (1990) showed that there are substantial interdecadal variations on the number of hurricanes striking the Caribbean region.

An overview of the Caribbean tropical storm activity over the past few centuries is given by LUDLUM (1989), MILLÁS (1968) and READING (1989). LANDSEA et al. (1999) constructed and analyzed a time series of Caribbean hurricanes based upon the presence of hurricanes within the Caribbean Sea and hurricanes directly affecting the landmasses, which surround the Caribbean. The region has shown dramatic changes in hurricane activity – averaging around 1.5 per year during the 1940s through the 1960s dropping to near 0.5 per year in the 1970s to the early 1990s. This is followed by the unprecedented (in this

---

[7] IAS: Caribbean Sea, Gulf of Mexico, Straits of Florida and adjacent seas.

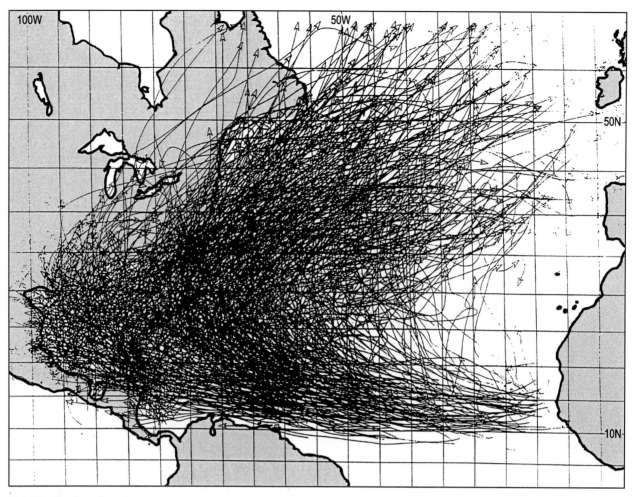

Fig. 51: Tracks of tropical storms over the North Atlantic Area from 1886 to 1995. In total 956 storms were recorded (METEOROLOGICAL SERVICE OF THE NETHERLANDS ANTILLES AND ARUBA, 1998). The density of the tracks illustrates the high frequency of hurricanes in this geographical region. During major hurricanes (categories 3, 4 or 5 on the SAFFIR-SIMPSON Hurricane Scale) like *Gilbert* (1988), *Hugo* (1989) and *Luis* (1995) wind speeds exceeding 300 km/h were observed.

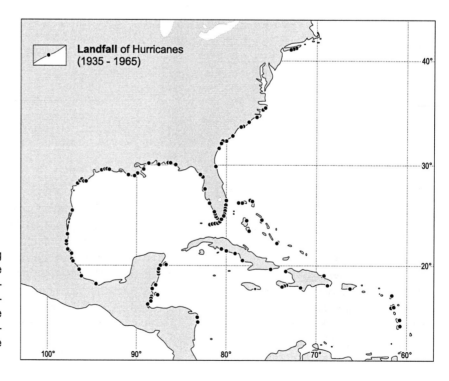

Fig. 52: The landfall of hurricanes during the time period 1935-1965. The coastlines along the northern Caribbean, the Golf of Mexico and Florida are more affected by hurricane landfalls than the southern Caribbean, where landfalls are very rare (after: LANDSEA, 1993).

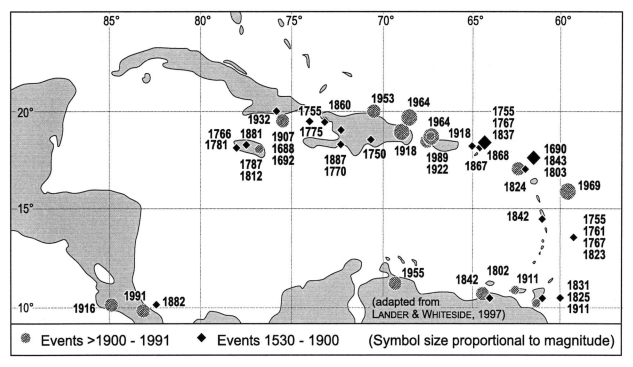

Fig. 53: Historical tsunamigenic events (46) in the Caribbean, not all events are present because 11 events lack their source coordinates (SOURCE: LANDER & WHITESIDE, 1997). The size of the symbol is proportional to the event's magnitude.

five decade time series) six hurricanes afflicting the region in 1996. (e.g. Central America, Cuba, Jamaica, Hispaniola, Puerto Rico, the Lesser Antilles, northern Venezuela and northern Colombia). The higher frequency of hurricanes compared to tsunami events is illustrated in Figure 52, which overviews the landfall of hurricanes during the time period 1935-1965.

Tsunami record

Although the Caribbean Sea region is well known for its hurricane hazard, the awareness of a potential risk for disastrous tsunamis was heightened only in the past few years. This is striking concerning the fact that practically all known causes of tsunamis generating mechanisms are present in the Caribbean. These include primarily earthquakes, but also landslides or mass flows, volcanic eruptions either submarine or subaerial, collapses of volcanic edifices, and teletsunamis (Fig. 53).

The compilation of a tsunami catalog over the last 500 years (LANDER & WHITESIDE, 1997) have highlighted the significant hazard in the Caribbean (Table 10). Eighty-eight tsunamis have been reported for the entire Caribbean region in the past 500 years, including 14 tsunamis reported from Puerto Rico and the Virgin Islands. 30 tsunamis caused significant damage including reports of as many as 9600 fatalities. The record for the last hundred years lists 20 tsunamis or about one every five years (LANDER & WHITESIDE, 1997). 1922 deaths are confirmed as being specifically related to tsunamis during the last 150 years.

This is a similar fatality level in the Caribbean as in Hawaii, Alaska and the west coast of the United States combined (LANDER & WHITESIDE, 1997).

Regarding the Caribbean catalog compiled by LANDER & WHITESIDE (1997), in no case observations of geomorphic tsunami evidence have been reported, but we suggest that at least three tsunamis might have been capable of boulder transport:

- The 1530 tsunami with a runup of 7.3 m near the coast of Paria and Cumana and near the Island of Cubagua.

- The 1867 Virgin Island tsunami, which generated waves with 9 m runup at St. Croix. This tsunami was observed throughout the eastern Caribbean causing damage as far south as St. Georges Harbor in Grenada.

- And least the 1918 Virgin Island tsunami with wave heights up to 6.1 m in Pt. Agujereada, Puerto Rico.

A catalogue of historical tsunamis for the Venezuelan Caribbean Coast reported in SCHUBERT (1994) lists additionally in 1726 a large wave at the Araya Península, which should have led to the partial destruction of a Spanish fort and a 10 m wave in 1906 at Macuto, Puerto Tuy. FERNANDEZ et al. (2000) compiled a tsunami catalog for Central America containing 49 tsunamis for the period 1539-1996, thirty-seven of them are in the Pacific and twelve are reported having affected the Caribbean coastline of Central America. They suggest that on the Caribbean side the Golfo de

Tab. 10: Historical Caribbean Tsunami Record from 1530 to 1991 (LANDER & WHITESIDE, 1997).

ORIGIN DATA			TSUNAMI DATA	
Date	Area	Location of Effects	Runup [m]	Comments
1530 09 01	Venezuela	Paria Cumana Cubagua	7.3	Sea rose 7.3 m and sank again near coast of Paria and at Cumana and near Island of Cubagua. Ground opened emitting black salt water and asphalt. Mountain at the side of the Gulf of Cariaco was cleft (earthquake). A fort and many houses destroyed, but not clear whether due to the wave, the earthquake or both.
1543	Venezuela	Venezuela		Waves noted. City of Cumana destroyed by earthquake?
1688 03 01	Jamaica	Port Royal, Jamaica		Shocks felt throughout the island and waves damaged ships in Port Royal. A ship at sea was damaged by a hurricane. No hurricane was reported.
1690 04 16 [Gregorian]	Leeward Is.	Charlotte Amalie, Virgin Islands Charleston, Nevis		The sea withdrew from Charlotte Amalie, St. Thomas, 16.5-18.5 m. Robson gives the date as April 5 for Nevis, but this is the Julian date. Earthquake of intensity IX caused landslides on volcanic Nevis Peak which caused the sea to withdraw 201 m from Charleston before returning in 2 mins.
1692 06 07 [11:40 Local Time LT]	Jamaica	Port Royal, Jamaica		Earthquake and subsidence destroyed 90% of the buildings in the city. Ships overturned, a frigate washed over tops of buildings. A wave 6 ft high traversed the bay. Along the coast of Liganee (possibly Liguanea Plain) the sea withdrew 183 or 274 m, exposing the bottom; upon returning the water overflowed the greater part of the shore. At Yallhouse (possibly Yallahs) the sea is said to have retired about 1.6 km. At Saint Anns Bay a large wave was reported. 2000 people killed by the earthquake and tsunami.
1751 10 18 19:00	Haiti Santo Domingo	Azua de Campostela, Haiti, Santo Domingo & El Seybo		The city of Azua de Compostela was destroyed by an earthquake and overwhelmed by the resulting tsunami. Santo Domingo also reported damaging waves at Santo Domingo and El Seybo.
1755 11 01 [10:24 Universal Time UT] Lat 36.0 N Long 11.0 W	Lisbon, Portugal	Saba	7.0	Teletsunami from the Lisbon (Portugal) earthquake. Waves of amplitude 7 m were observed at Saba, 3.6 m at Antigua and Dominica, 4.5 m at St. Martin, leaving a sloop anchored in 4.6 m of water was left laying broadside on the dry bottom, 1.5-1.8 m at Barbados, where the wave had a period of 5 minutes and the water was black as ink. This could be a local landslide tsunami or seiche triggered by the Lisbon wave. At Martinique, at some places the water was reported to have withdrawn for 1.6 km and at other places it flowed into the upper level rooms of the houses. The lowlands on most of the other French Islands were inundated. There is a report of Santiago de Cuba being nearly inundated in 1755 but the month and day were not given. This is probably from the Lisbon tsunami.
		St. Martin	4.5	
		Antigua & Dominica	3.6	
		Barbados	1.5-1.8	
		Martinique		
		Santiago de Cuba		
1761 03 31 [13:14 UT]	Lisbon, Portuga	Barbados		An earthquake near Lisbon, Portugal, caused an extraordinary flux and reflux of the sea at Barbados.
1766 06 11 [4:05 UT]	Cuba	Jamaica		An earthquake lasting 1-1/2 to 7 minutes hit Santiago de Cuba and Bayamo. Ships at sea 7.2 km from the coast of Jamaica rolled so much that their gunwales were immersed in the water. Ships in deep water would not experience a tsunami. Either the ships were near the coast or in shoaling water or the wave was a storm wave but no storm was reported.
1766 10 21 [9:00 UT]	Venezuela	Cumana, Venezuela		Very violent shocks raised Cumana and caused the island of Orinoco to sink and disappear. In many places the water surface was disturbed. This is a possible tsunami report.
1767 04 24 [6:00 UT]	Martinique, Barbadoo	Martinique, Barbadoo		The sea was much agitated and ebbed and flowed in an unusual way.
1770 06 03 [19:15]	Haiti	Golfe de la Gonave, Haiti		Waves noted at Golfe de la Gonave and Archaie. La Saline Mountain foot partly submerged. The sea inundated 7.2 km inland.
1775 03	Haiti	Hispaniola		Three strong shocks were felt on Hispaniola. Several store houses were destroyed and great damage was done by the sea.
1775 12 18	Hispanola Cuba	Hispaniola Cuba		Three earthquakes reported and waves did extensive damage. Rubio does not mention any effects in Cuba.
1780 10 03 [22:00 LT]	Jamaica	Savanna la Mar, Jamaica	3.0	An earthquake occurred during a hurricane at 22:00 LT The sea rose to a height of 3 m at 0.8 from the beach and swept away a number of houses. Ten people were killed by the wave and about 300 more by the storm. All vessels in the bay were dashed to pieces or driven onshore. Waves may have been storm surge.
1781 09 01 [14:20 LT]	Jamaica	Jamaica		In 1781 a series of waves and disastrous earthquakes nearly ruined the Island. There are no reports of earthquakes for this day, but there are reports of a major hurricane on Aug I at 14:20 LT.
1787 10 27	Jamaica	Montego Bay, Jamaica		A small local shock was felt at Montego Bay and the vessels in the harbor were agitated. Mallet reports earthquakes in Jamaica on Oct 1 and 21 at Kingston and Port Royal. This would be a low validity report as no wave was cited and the agitation may have been a report of a seaquake effect.
1802 03 19	Leeward Is.	Antigua St. Christopher		Earthquakes were reported in February and March with the largest on this date. It was accompanied by great agitation of the sea. Intensity IV.
1802 08 05	Venezuela	Orinoco River, Venezuela		Earthquakes at Cumana caused the water of the Orinoco River to rise so high as to leave part of the bed dry. This could describe wave action near the mouth of the river, or bore action. The rudder of a vessel was broken.

Chapter 3: Hurricanes and Tsunamis

Tab. 10: continuation

Date	Area	Location of Effects	Runup [m]	Comments
1812 11 11 [10:50 UT]	Jamaica	Jamaica		The sea was much agitated following an earthquake. This could describe wave action or seaquake action.
1823 11 30 [2:45 LT]	Martinique	Saint-Pierre Harbor, Martinique		At 2:45 LT a strong undulation (earthquake) was followed by a tidal wave at 3:10 p.m. which caused some damage in Saint-Pierre Harbor.
1824 09 13	Guadeloupe	Plymouth, Montserrat		Earthquakes were felt at Basse Terre on the 9th and on the 13th there was a remarkable rise and fall of the tide at Plymouth Montserrat. There had been a terrible storm and heavy rain on September 7-9.
1824 11 30	Martinique	St. Pierre		Severe shock at St. Pierre. A very high tide threw many ships upon the strand. Heavy rain followed lasting 10 days.
1825 09 20	British Guiana	Demerara County, British Guiana		Local earthquake and oscillations of the sea were noted. An earthquake (MMI=VIII) was also noted at Trinidad Tobago St. Vincent and Barbados.
1831 12 03	Trinidad & St. Christopher	Trinidad St. Christopher		An earthquake occurred. The sea was in a state of violent agitation. Note the large distance between reporting areas. An earthquake was also reported at Grenada, St. Vincent, and British Guiana.
1837 08 02	Martinique			Several shocks accompanied by a large wave occurring during a hurricane. Source of wave uncertain.
1842 05 07 [17:30 LT]	Guadeloupe	Guadeloupe: Basse Terre,	0.9	A strong earthquake (Mw=8.3) at 17:30 LT produced waves with heights reported; a wave carried away all floatable objects at Deshaies and Sainte Rose; at Gouyave, Grenada (Charlotte Town), there was some damage; at Haiti, a destructive tsunami struck the north coast; at Mole Saint-Nicolas, Cap Haitien, there was extensive destruction caused by the earthquake and tsunami; at Port-de-Paix the sea receded 60m and the returning wave covered the city with 5 m of water. About 200-300 of the city's 3,000 inhabitants were killed by the earthquake and tsunami. It was observed at Fort Liberte, Mole St. Nicolas, and Santiago de los Caballeros. At Hispaniola there was destruction on north coast by waves of 2 m. Note the large area of this event which suggests a teletsunami, but the earthquake was felt at Haiti, Jamaica, Puerto Rico, and other islands. Note also the missing locations such as Puerto Rico for which no tsunami report is available.
		Deshaies, & Sainte Rose	8.3	
		Bequia Island	1.8	
		St. Johns, Virgin Is.	3.1	
		Charlotte Town, Grenada, Haiti: Cap Haitien, Port-de-Paix, Fort Liberte, Mole St. Nicolas, Santiago de los Caballeros, Hispaniola	2.0	
1843 02 08	Antigua	Antigua	1.2	An earthquake was felt at Pointe-a-Pitre, Guadeloupe, St. Lucia, St. Kitts, Montserrat, Martinique, and other islands. The sea rose 1.2 m but sank again immediately. A volcanic eruption near Marie Galante followed on Feb 17, which ejected jets and columns of water and probably resulted in a minor tsunami.
1852 07 17	Cuba	Santiago de Cuba		At Santiago de Cuba the bay was affected by a strong surge which affected the port buildings and loading docks. It probably was the product of an earthquake which affected the US frigate Tropic which was about 112 km from Jamaica.
1853 07 15	Venezuela	Cumana, Venezuela		A violent earthquake (MMI=X) in Cumana followed by a tsunami.
1856 08 04	Honduras	Honduras	5.0	An earthquake and 5 m wave. Earlier wave reported some years earlier. Not known if on the Caribbean or Pacific coast.
1860 03 08	Hispaniola	Hispaniola: Golfe de la Gonaves, Cayes, Acquin, Anse-a-Veau.		An earthquake (Ms=7.5) was reported from Port-au-Prince and Anse-a-Veau. Waves were reported from Golfe de la Gonaves, Cayes, and Acquin. At Anse-a-Veau the sea withdrew and broke with acrash on the shore.
1867 11 18	St. Thomas, Virgin Is.	St. Thomas, Charlotte Amalie, Altona	6.0	An earthquake (Ms=7.5) occurred in the Virgin Is. Waves around about 15 mins after the earthquake. At Charlotte Amalie, the height was 2.4 m above the sealevel at the wharf, and the lower part of city was flooded. The water receded nearly 100 m and returned as a wave 4.5-6 m high swamping small boats in the harbor. The wave penetrated 76 m inland. The USS De Soto was damaged, 11-12 people were killed. At Altona, houses were washed far inland, and there was some damage at Hassel I. At Christensted, St. Croix, waves swept inland 91 m, and at Gallows Bay, 20 houses were damaged. At Frederiksted the sea withdrew and returned as a wall of water 7.6 m high leaving the USS Monongahela stranded. Five were killed, 3-4 injured, and 20 houses were damaged. At Puerto Rico, at San Juan, the river water rose 0.9-1.5 m and at Vieques, high waves were observed. At Fajardo, a very small wave was reported, and at Yabucoa the sea retreated and inundated l 37m on its return. In the British Virgin Islands, at Peter Is., a wave was noted and people fled to Tortola. At Roadtown, Tortola, a 1.5 m wave swept some houses away. At Saba, there was some damage. At St. Christopher, the wave was also observed. At St. Martin and St. Barthelemy, there was some damage. At St. Johns, Antigua, the wave had a height of 3.0 m. At Basse Terrre, Guadeloupe, the height was 1.0 m with the sea retreated far from coast. At Deshailes, houses in village were destroyed. At Isles des Saintes, there was a slight swell, and at Fond du Cure, houses inundated to a depth of 1 m. At Pointe-a-Pitre, there was a slight swell, and at Sainte-Rose, a 10
		St. Croix, Friedricksted,	7.6	
		Christensted / Gallows Bay	1.5	
		Puerto Rico, San Juan,	1.5	
		Fajardo, Yabucoa / Vieques	1.5	
		British Virgin Is: Tortola, Road Town, Peter Is., Saba, St. Christopher, St. Martin / St. Barthelemy, St. Johns, Antigua	3.0	
		Guadeloupe: Basse-Terre,	1.0	

Tab. 10: continuation

Date	Area	Location of Effects	Runup [m]	Comments
		Deshailes, Isles des Saintes,	10.0	m wave. The sea withdrew 100 m and flooded and damaged houses on return. It was observed at Martinique and St. Vincent had unusually high water. At Grenada, Gouyave (Charlotte Town), the height was 3 m and at St. George, 1.5 m. At Becquia Is., it was 1.8 m.
		Fond du Cure, Pointe-a-Pitre / Sainte-Rose Martinique, St.Vincent Grenada, Charlotte Town,	3.0	
		Becouia Is.	1.8	
1868 03 17	Puerto Rico	Puerto Rico, Arroyo / Naguabo, St.Thomas, Charlotte Amalie	0.6	An earthquake and tsunami were observed at Arroyo and Naguabo. At St. Thomas, Charlotte Amalie, it was .6 m, with a small recession and flooding.
1874 03 11 [4:30 LT]	Lesser Antilles	Dominica St. Thomas, Virgin Is.		A submarine shock to the southeast of St. Thomas shook the island and ships in the harbor. Simultaneously the water in the bay, then perfectly still, appeared turbid as though clouded by sand and mud. A little later strong ripples from the south agitated the water surface lasting some time. This probably was the tsunami and the earlier effects from the seismic waves agitating the bottom. At Dominica, the steamer Corsica reported at 5:00 LT a series of heavy rollers in the harbor lasting half an hour and rendering communication with the shore impossible. They did not feel the earthquake. The reduced effects at Charlotte Amalie may indicate a source on the eastern side of the island.
1918 10 11	Puerto Rico	Aguadilla	2.4 -3.3	A magnitude 7.5 earthquake caused a wave of 2.4-3.3 m above sealevel at Aguadilla, which destroyed 300 huts and drowned 34 people. At Cayo Cardona, water rose 75 cm on the west side of the island. At El Boqueron, the wave dropped 1.5 m and rose 90 cm above mean sealevel. About 800 m southeast near the entrance to the bay the water rose only 45 cm. At Punta Borinquen Lighthouse, the wave was 4.5 m above sealevel. In a low area just southwest of the lighthouse thewave penetrated 91 m inland. Submarine cables were cut in several places. At Gaunica, 45 cm waves were observed. At Isabella, the water rose 1.8 m. At Isla Caja de Muertos, water rose 1.5 m covering 15 m of the beach. At Isla Mona, the receding water bared the reef and the returning wave was 3.6 m above sealevel washing a pier away, and flooding a cistern. At Mayaguez, a wave entered the first floors of buildings near the waterfront and destroyed a few native huts and a brick wall was overturned. Water levels reached 40 to 150 cm above sealevel. At Playa Ponce, slight water movements were observed. At Puerto Arecido, waves 30 to 60 cm high were observed and a bore about 10 cm went up the Rio Grande. At Punta Agujereada, waves estimated at 5.5 to 6 m uprooted several hundred palm trees and destroyed several small houses. Eight people drowned. At Punta Higuero Lighthouse, wavesuprooted coconut palms and crossed railroad tracks 4.9 m above sealevel while 800 m southeast of the lighthouse the water rose 2.6 to 2.7 m. At Rio Culebrinas, 1000 kg blocks of limestone were moved 46 to 76 m slightly downhill. Waves were at least 3.7 m high. At Rio Grande de Lioza, water receded and rose about 90 cm. At St. Thomas, Virgin Islands, Charlotte Amalie, the water rose 45 cm and at Krum Bay, 1.2 m. At Santo Domingo, Hispaniola, water of the Rio Ozama fell and rose 60 cm with a period of 40 mins. Waves were noted at Tortola.
		Isabela	1.8	
		Cayo Cardona	0.75	
		El Bouqueron	2.4	
		Punta Borinquen	4.3	
		Isla Caja de Muertos Gaunica	1.5	
		Isla Mona	0.6	
		Mayaguez	3.6	
		Puerto Arecido	1.5	
		Punta Agujereada	0.6	
		Punta Higuero	6.0	
		Rio Culebrinas	4.9	
		St. Thomas: Charlotte Amalie / Krum Bay	3.7 0.45 1.2	
		Santo Domingo, Hispaniola	0.6	
		Tortola		
1918 10 25 [3:42 UT]	Puerto Rico	Puerto Rico, Mona Passage		Submarine cables were cut again and a steamer rolled heavily. Waves were recorded on the tide gauge at Galveston Texas.
1922 05 02	Puerto Rico			A wave was recorded on the Galveston, Texas, gauge which has been associated with a small earthquake in Vieques Is., but the small earthquake does not seem likely to have produced a recordable tsunami.
1929 01 17	Venezuela	Cumana, Venezuela		City was destroyed by an earthquake (Ms=6.9) and a steamer off shore was endangered by a huge wave. The tidal wave following the earthquake caused much damage. Many sailboats were wrecked.
1931 12 01	Cuba	Cuba: Playa Panchita, Rancho Veloz, Las Villas		At Playa Panchita, Rancho Veloz, and Las Villas, waves beat on the beaches. Coastal houses were inundated damaging household furniture. A hurricane passed north of Cuba on Nov 25, and became a tropical storm in Cuba by Nov 30. No earthquake resorted.
1932 02 03 [9:16 UT]	Cuba	Santiago de Cuba		Small waves were reported at the time of an earthquake at Santiago de Cuba by a north American boat. A strong earthquake affected 80% of the buildings in this city, but the wave was of little importance.
1939 08 08 [3:52 UT]	Cuba	Frances Cay, Cuba		Frances Cay reported movement of the sea that woke up the sailors of a boat. It was a II on the Rudolph scale. The earthquake (Mb=5.6) that caused this tsunami affected the north of the province of Las Villas and the epicenter was located in the sea. ISS places the epicenter within the island of Cuba.

Tab. 10: continuation

Date	Area	Location of Effects	Runup [m]	Comments
1946 08 04	Dominican Republic	Matanzas, Julia Molina & Samana, San Juan, Puerto Rico, Bermuda. Daytona Beach, Florida, Atlantic City New Jersey	2.4	The town was severely damaged and 100 people killed although the wave probably was only 2.4m. At Villa Julia Molina the wave was estimated to be 3.6 to 4.6m high but caused little damage. At Cabo Samana several ebbs and flows were observed. It was recorded at San Juan, Puerto Rico, 36 minutes after the earthquake. It was also recorded at Bermuda at 2:07 after the earthquake and at Daytona Beach, 3:59 and Atlantic City, 4:49.
1953 05 31 [19:58 UT]	Dominican Republic	Puerto Plata, Dominican Republic	0.6	Recorded on the Puerto Plata tide gauge at 6 cm height. May be a wave from hurricane Alice which was in the area at this time.
1955 01 18	Venezuela	La Vela, Venezuela		A wave was reported; four ships were wrecked and four waterfront buildings damaged. An earthquake (Mb=5.5) off the coast of Panama is listed for this time.
1968 09 20 [6:09 UT]	Venezuela			A report of a tsunami has not been verified. Hurricane Edna was passing north of Venezuela at this time and an earthquake (Ms=6.2) occurred near the coast of Venezuela.
1969 12 25 [21:32 UT]	Leeward Is.	Barbados, Antigua & Dominica	0.46	Recorded at Barbados, Antigua, and Dominica with a maximum amplitude of 46 cm at Barbados.
1985 03 16 [14:54 UT]	Leeward Is.	Basse Terre, Guadeloupe		A moderate earthquake (Ms=6.3) caused damage at Montserrat. It was also felt at Antigua, St. Kitts, and Puerto Rico. A several cm tsunami was recorded at Basse Terre, Guadeloupe.
1989 11 01 [10:25 UT]	Puerto Rico	Cabo Rojo, Puerto Rico		A small tsunami was reported from an earthquake (Mb=5.2).
1991 04 22 [21:56 UT]	Costa Rica	Panama: Bokas del Toro, Carenero Island, San Cristobal Island, Puertobelo, Colon	0.10 0.60	At Bokas del Toro, Panama, people reported that Las Delicias sand bank normally covered by 60 to 90cm of water emerged as the sea receded less than ten minutes after the earthquake and remained above water for five to seven minutes. Afterwards several waves entered the bay with great force flooding 50 to 100 m in the flat northern part of the town. At Carenero Island violent waves destroyed dwellings. At San Cristobal Island the sea receded several meters for about 45 minutes. People went on the beach to catch trapped fish. It was also observed at Bastimento, Cristobal, 10 cm, Puertobelo, W. Panama, 60 cm and recorded at Colon.

Honduras-zone and the coasts of Panama and Costa Rica have a major hazard risk.

External sources might also generate significant tsunamis in the Caribbean. The Lisbon earthquake of 1755 caused 2-3 m waves in Antigua and Barbados, 7 m waves in Saba, and 4 m waves in St. Maarten. Presently, there is concern over the source of the next major Caribbean tsunami center at the submarine volcano Kick'em Jenny, located 8 km off the north coast of Grenada. The volcano is the most active in the Lesser Antilles with at least 11 eruptions in historical time. SMITH & SHEPHERD (1993) estimated wave run ups for several types of energy release, with a "worst case scenario" equivalent to the Krakatao-eruption.

LANDER & WHITESIDE (1997) noted, that in general there is often confusion in historical reports between a true tsunami and local effects caused by hurricanes. Therefore, the Caribbean tsunami catalog might probably be contaminated with a number of reported tsunamis, which were actually sea waves caused by hurricanes. On the other hand the list of tsunami effects might show gaps in reporting. For example none of the Caribbean tsunamis have generated damaging effects at teletsunami distances, but some have been recorded as far as on the East coast of the United States. Another striking feature is that since the beginning of the tsunami record in 1530 the ABC-islands are not listed in any tsunami catalogue, but the north coast of Venezuela (in particular the region around Cumana) is mentioned for the years 1530, 1543, 1766, 1802, 1853, 1929, 1955, and 1968, and e.g. the Virgin Island Tsunami from 1867 reached Curaçao after 90 minutes, but no damage was reported.

Therefore, the importance of geomorphologic research of paleotsunami studies can be highlighted in order to differentiate clearly between storm and tsunami deposits in order to complete the historical tsunami record for the Caribbean.

Field evidence of supratidal debris deposition
Debris deposits through hurricane-generated waves are described by HERNANDEZ-AVILA et al. (1977) for the south coast of Grand Cayman and Puerto Rico. The size of the deposits on Grand Cayman is on average 58 × 20 × 10 cm with some fragments reaching 60-100 cm in diameter. In Puerto Rico a rampart formation is accumulated with an average height of 1.5 m and a length of 50 m, with boulder sizes from 3 cm to 2.5 m. The authors demonstrate via field experiments that hurricane waves are capable of depositing the debris formation with an *Acropora palmata* reef located 300 m offshore in 8-10 m depth acting as the source of the debris. One of the most

intense field surveys of effects of hurricanes on reefs was carried out by STODDART (1962), who documented in detail the impact of hurricane *Hattie* (1961), one of the most destructive in Honduras. However, the geomorphologic evidences for debris deposition are restricted to the accumulation of coral rubble in minor quantities.

One of the few published reports of geomorphologic evidence for a tsunami impact was given by SCHUBERT (1994), who described a coral gravel located on a 10-20 m high erosional terrace, west of Puerto Colombia (north-central Venezuelan coast) with an average uranium-series age of 1,300 ± 160 a BP. These deposits were interpreted as evidence of a prehistoric strong tsunami with a runup height of about 15 m. Large coral gravel deposits have been also reported from the Venezuelan offshore islands (SCHUBERT & VALASTRO, 1976). They are particular well-developed along the southern margin of Los Roques, where several ridges up to 5-10 m high form a barrier along the beach, and on the island of La Orchila. No detailed studies have been done to date the deposits or to find evidence of their origin. WEISS (1979) described another possible tsunami evidence on the Caribbean Coast of Venezuela for a saline lagoon at the island of Cayo Sal, where the sedimentary succession in the lagoon was overturned. Radiocarbon dates suggest that this event took place between 770 and 500 BP. Recently sedimentary evidence for at least two tsunamis in Puerto Rico with a runup of about 6 m has been documented (MOYA, 1999).

Boulder deposits are hitherto only described for the coast of Grand Cayman (JONES & HUNTER, 1992) and for north Eleuthera island, Bahamas (HEARTY, 1997). On the rocky coastline of Grand Cayman two stretches of boulder accumulations occur, which are irregularly distributed, reaching as far as 100 m inland from the present coastline (JONES & HUNTER, 1992). The boulders are up to 5.5 m long, 3.4 m wide and 2.3 m high with an estimated weight of up to 40 tons. Some boulders of 10 t were moved up to 18 m vertically and 50-60 m horizontally. Encrustations of vermetids and borings of sponges and *Lithophaga* show that they must have been submerged prior to their present position. Radiocarbon dating of a coral species (*Astrangia sol.*) yielded an age of 1662 AD (1625-1688 AD at 68% confidence limits). The authors suggest, that these boulders may have been moved by hurricane-generated waves or a tsunami that was triggered by an earthquake or slumping on submarine slopes. A Pleistocene tsunami event is according to HEARTY (1997) a plausible cause for the deposition of seven boulders measuring 100-1000 m³ along the coastal ridge of north Eleuthera. Some are situated on the ridge crest up to 20 m asl. The boulders were deposited probably during oxygen-isotope stage 5e or 5d as shown by their stratigraphic setting and by amino acid racemization ratios. He stated that the largest boulder is about 10 times the size of the largest Holocene ones moved by waves in the area and tsunamis are a reasonably possibility as a transporting mechanism. Considering a storm as a cause, the storms during the last interglacial must have been of much greater intensity than those occurring in the region during the late Holocene.

3.3.4 Hurricanes and Tsunamis on Curaçao, Bonaire and Aruba

Hurricane record

Curaçao, Bonaire and Aruba are located on the southern border of the hurricane belt. Once every 100 years considerable damage is experienced by tropical cyclones passing over or just south of the islands, and on average once every 4 years a tropical cyclone occurs within a radius of approx. 100 nm, but mostly passing to the north without causing serious bad weather (METEOROLOGICAL SERVICE OF THE NETHERLANDS ANTILLES AND ARUBA, 1998). During the time period 1605-2000 in total 14 hurricanes and 19 tropical storms, with maximum wind velocities between 100-120 mph (= 180-210 km/h) near the center, passed the islands within the 100 nm zone (Fig. 54). Whereas e.g. on the Windward Antilles the island of St. Eustatius (17.5°N) experienced in total 192 severe hurricanes with wind velocities up to 250 km/h.

The damage to the infrastructure of the islands is mainly due to flash floods from associated heavy rainfall. Even major hurricane *Hazel* (1954) with a close track approximately only 90 km to the north caused the most damage by rainfall (Aruba 250 mm). *Hazel* recorded max. sustained winds near the center of 190 km/h and max. winds of 50 km/h with gusts to 90 km/h on the islands. A review of the historical colonial reports and newspaper articles published after the hurricane events of 1784, 1831, 1877, 1886 and 1892 brought no evidence for any destructive impact or confirmations for extraordinary storm depositions along the coastlines of the islands (Centraal Historisch Archief, Curaçao: "The Gentleman's Magazine" 1785, Vol. 57, p. 154; Colonial Report 1877, 1878, 1887; "Curaçaosche Courant" from June 1831, August 20 and 27, 1886, October 14, 1892).

The most significant event in the past few years was hurricane *Lenny* (1999), an extremely rare hurricane with wind speeds >160 km/h, formed south of Jamaica and moved eastward toward the Lesser Antilles (Fig. 55). This direction of travel for a sustained period, is the first reported in the entire 113 year

Chapter 3: Hurricanes and Tsunamis

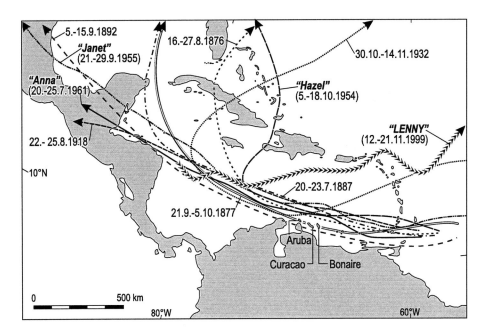

Fig. 54:
Only few hurricanes passed within 100 nm from Curaçao, Bonaire and Aruba over the time period from 1605 to 1998.

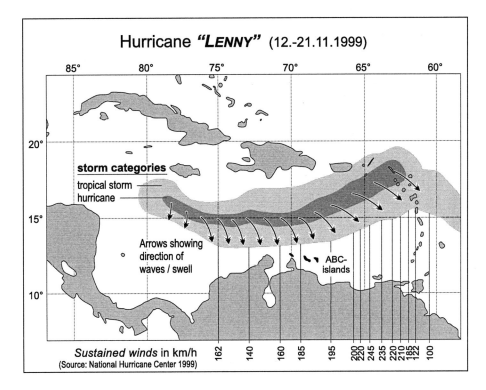

Fig. 55:
Hurricane *Lenny*'s pass through the Caribbean. The system was formed south of Jamaica and moved eastwards towards the Lesser Antilles. The ABC-islands experienced heavy surf due to stronger wind velocities and wave height on the right side of its track (after: GUINEY, 2000).

hurricane record (GUINEY, 2000). When the storm was approximately 240 km to the southwest of Jamaica, the wind speed and central pressure reached significant values allowing to classify the system as a hurricane. The maximum intensity occurred when *Lenny* was just south of St. Croix, here the system was just six miles per hour short of being classified as a Category 5 hurricane on the SAFFIR-SIMPSON Hurricane Scale (GUINEY, 2000).

The islands of Aruba, Bonaire and Curaçao all experienced heavy surf conditions along their southwestern coastlines as *Lenny* passed 250-500 km north of the islands. The waves varied along the coasts, but were reported to be mostly in the rage of 3-6 m (pers. commun. A. DEBROT, 2001). During the period between November 15th to 17th, 1999, swells caused severe damage to the coral reef and infrastructures like small vessels and beach structures.

Two other events with similar unusual wave heights and rough seas were related to the tropical storms *Joan* in 1988 and *Bret* in 1993. During *Joan* maximum observed sustained winds were confined to 65 km/h with gusts to 90 km/h and wave heights of 2-3 m were observed (KOBLUK & LYSENKO,

1992). During *Bret* on all three islands wind gusts over 75 km/h were recorded (METEOROLOGICAL SERVICE OF THE NETHERLANDS ANTILLES AND ARUBA, 1998).

Field evidence of supratidal debris deposition

The debris deposits of the Netherlands Antilles and Aruba were hitherto exclusively attributed to hurricane-generated waves (DE BUISONJÉ, 1974). In the literature, hurricane *Janet* (1955) is mentioned as a possible cause for the rampart formations along the windward coasts of the ABC-islands. But considering that the region associated with the highest wind velocities of up to 278 km/h was located approximately 600 km north and 1500 km west of the islands, it seems to be extremely unlikely that *Janet* acted as a source mechanism, in particular because the weathering state of the debris indicate much older ages for the fragments. Additionally, the occurrence of mining activities, which reach back into the last century, show that the debris have been existed already before 1955, the year of hurricane *Janet*.

So far the occurrence of tsunamis has been completely neglected for the Netherlands Antilles. This study will provide a detailed description and documentation of the boulder deposits and rampart formations of the Leeward Netherlands Antilles and discuss the source mechanism for different depositional units with paleotsunami events as the most realistic probability for the greater part of the debris deposits.

4. Materials and Methods

The basic activity of paleotsunami studies is the identification, mapping, correlation and dating of tsunami deposits. The fundamental goal is to reconstruct past and pre-historic tsunamis. Ideally, these reconstructions would include quantified estimates of tsunami size and extent, a determination of the recurrence interval and the specification of the tsunami-generating event and origin.

Key problems in paleotsunami studies include the very basic issue of the positive identification of such deposits: Can we distinguish them from storm and flood deposits? Recent studies of historic tsunami deposits have contributed to the understanding of such deposits and some diagnostic criteria have been developed for identifying paleotsunami deposits (CHAGUÉ-GOFF & GOFF, 1999). However, most of these criteria are developed for fine sediments and are lacking for more coarse and boulder size material. Some characteristics for the identification of such tsunami debris deposits on the ABC-islands will be presented and discussed in detail in Chapter 5.

Another key problem in paleotsunami studies is the dating and the correlation of events: When comparing one locality to another, are we looking at the same tsunami deposit or at deposits from separate events?

The standard dating technique for Holocene and Late Pleistocene marine carbonates is the radiocarbon method. Nevertheless, the sampling of appropriate material in debris deposits can be problematic, as older reworked or contaminated material can be incorporated. A list of criteria for the selection of the debris material was defined in order to minimize errors in the dating routine. For fine sediments optical dating (OSL) is suggested to be the best method available, assuming that the sediments were exposed to daylight during the sedimentation process (CHAGUÉ-GOFF & GOFF, 1999).

The third key problem is the quantification of paleotsunami events – how well can wave size or paleo-run up be calculated? A few studies have tried to answer these questions (HARBITZ, 1991; DAWSON & SMITH, 2000), but still paleotsunami models remain tentative.

4.1 Field Survey

The fieldwork has been carried out from June 2000 to January 2001. During this time, D. Kelletat joined in the field investigations from the 10-28 January 2001 and for a final review from 18-31 August 2001.

4.2 Characterization System

Based on the surveys a characterization system for the debris deposits was developed. Grouping of the units, distinguished on aerial photographs and the field survey, was conducted with regard to similarity in composition, morphology and their position in heights and distance from the coastline. In the classification distinction was made between:

1. Debris ridges with direct contact to coastline and surf zone (recent or subrecent).
2. Debris ridges located several meters above sealevel and in distance to the coastline (subrecent).
3. Ramparts (subrecent).
4. Single tsunami boulders and tsunami boulder assemblages (subrecent).

The units will be described and discussed in detail in Chapter 5.

4.3 Situmetry and Shape Measurements

One part of the fieldwork concentrated on the measurement of tsunami boulders regarding to size, heights and the distance from the coastline. The boulders mostly consist of Pleistocene calcareous limestone material of the Lower Terrace. They are ranging in size from 1.0 × 1.0 × 1.0 m (minimum mapping size) to 6.5 × 4.6 × 2.6 m and in volume from 1-112 m^3. Considering a specific weight of 2.5 g/m^3 (dense limestone), values up to 281 tons in weigth can be calculated. Since most boulders are elongated a survey of their long axis orientation was carried out, but no clear distribution could be observed. Conversely, the ramparts show a remarkable narrow range of imbrication axis, which were determined with situmetric measurements according to POSER & HÖVERMANN (1952).

To determine the source of the tsunami debris we analyzed with a statistical approach the shape and the material of the fragments within representative areas of 16 m^2 in leeward and windward ridges and ramparts on Curaçao and Bonaire. Terrestrial origin could be excluded in all cases, since only marine carbonates are present. The method is in the first place for descriptive purposes, nevertheless the different shapes give evidence about their genetic origin and the environment they have been derived from. The amount of examined fragments per area ranges from 140 to more than 600 depending on their size. In total, seven classes, which could be identified easily in the field, have been distinguished:

1. Well preserved coral species
   The class contains coral fragments with mostly intact preservation of the coral skeleton microsculpture, no diagenetic alternation (100% Aragonite), no significant traces of reworking during or after transport or deposition processes. The origin of the fragments from foreshore reefal environments can certainly be confirmed as correct. All observations point to a younger age of the dislocation during one single event.

2. Bad preserved coral species
   The class characterizes fragments with significant alterations of the shape, but still the species are identifiable. The changes are due to abrasion or bioerosion in the foreshore environment or to weathering after deposition. No doubt, that these fragments derive from the subtidal as well as Class 1, but the kind and process of reworking, the age as well as the number of dislocations remain more unclear.

3. Well-rounded fragments
   The well-rounded shape of these class members is the result of their stay in the subtidal environment with the occurrence of abrasional processes due to wave action. If the roundness is very well preserved, time of dislocation may be young without following influences of terrestrial weathering.

4. Bioerosive sculpturing predominant
   This class contains a wider range of shapes, which owe their morphological expression different processes of bioerosion in the littoral environment: The subtidal, where boring is the predominant bioerosional process and the supratidal with grazing as the main factor of bioerosive activity. The perforations of coral or rubble fragments by boring organisms do not change the shape of the rubble significantly, whereas grazing activities in the supratidal will transform the exposed limestone surface in a rock pool zone with sharp protruding limestone peaks with concave morphologies. The debris derived from this environment reflect these morphology. In the later one boring holes are absent. By far most of the counted debris fragments in ridges show signs of borings, in contrast to the windward ramparts, where rock pool fragments are more common. Consequently, it can be concluded that the members of this class derive from the subtidal in ridges and mainly from the supratidal in ramparts.

5. Karstified fragments
   The class contains fragments whose surface is heavily altered by terrestrial karstification processes. The surface shows an irregular microtopography, and the original shape is often very difficult to recognize. Karstification certainly takes place after the deposition process, which may be linked to a higher age. Wide and dense boring holes and in particular the supratidal concave bioerosive shape of a fragment support the process, as a result by far most of the fragments in Class 5 may derive from the supratidal environment.

6. Breakage at one side
   Classes 6 and 7 contain information about the extent of damage or fragmentation, e.g. corals with less or heavy skeletal damage. This kind of damage can be identified at fragments of Classes 1, 2 and 3, in nearly all cases the fragments derive from the subtidal.

7. Breakage at several sides
   Contains coral fragments with gross skeletal damage resulting in the colony broken into two or more parts. The interpretation is same as Class 6 in interpretation of origin, although fragmentation may be made while breaking of in the subtidal, by transport or at the place of deposition by impacts of other coarse debris.

Summarizing we can distinguish three environments acting as source areas for the debris deposits (Fig. 56). For differentiation and identification of hurricane and tsunami debris deposits a catalogue of typical geomorphologic characteristics is formulated in Table 11. Fig. 57 visualizes the differences in the rigde formations.

4.4 Interpretation of Aerial Photography

The maps that constitute the study were all constructed from color vertical aerial pictures taken by HANSA-Luftbild in 1998 (Curaçao) and 1996 (Bonaire)[8] and low oblique color aerial photographs as well as video material taken from a single-engined high wing aircraft by us.

The aerial pictures have a scale of 1 : 4000 (Curaçao) and approximately 1 : 10 000 for Bonaire. For Aruba only our own low oblique photographs were used. The vertical aerial pictures were digitized with a Canon Scanner (CanoScan FB 12005) with a resolution of 300 dpi and georeferencend with the Image Analysis Extension of Arc View 3.0. The subsequent mapping of the hurricane and tsunami deposits was carried out with the software packages Arc View 3.0 and Corel Draw 9.0. In this way extensive areas along the coastlines of Aruba, Bonaire and Curaçao were mapped and verified through field observations.

---

[8] *The aerial pictures were kindly put at our disposal by the Dienst Openbare Werken, Afdeling Topografie in Curaçao and by the Department of Physical Planning, Environmental and Natural Resources Section of Bonaire.*

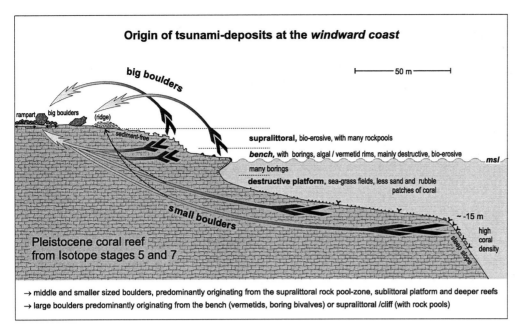

Fig. 56: Main tsunami depositional types and their genetic source in the coastal environment.

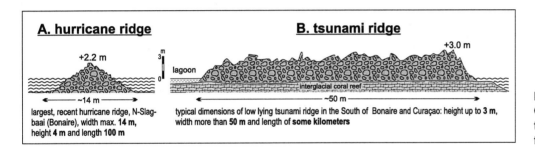

Fig. 57: Cross-section of a typical hurricane and tsunami ridge.

Tab. 11: Characteristics of hurricane and tsunami debris deposits.

	Hurricane debris	Tsunami debris
material	mostly corals	corals, Pleistocene limestone fragments and boulders
boulder size	small to medium sizes	medium to huge sizes
roundness	well rounded	less rounded
weight	weight rarely exceeds 1 t	weight up to >200 t
altitude [m asl]	0.5 - 1.5	0.5 - 15
distance to coastline	>5 m	20 - 400 m
height of deposit (max.)	2.5 m (ridge)	4 m (ridge)
width of deposit	<15 m	50 m (ridge) up to 400 m (ramparts)
micro-topography	sharp crest	wide crest, undulating topography
internal structure	chaotic	imbrication common

4.5 Relative and Absolute Age Dating of the Deposits

4.5.1 Relative Age Dating

A number of stratigraphical, morphological and historical data allow us to define the period in which the tsunami event should have occurred (KELLETAT & SCHELLMANN, 2001 a,b). These relative age indications include:

- Color of the weathered surface of the deposits,
- Intensity of limestone solution after deposition,
- State of weathering of fractured surface areas of the deposits,
- Conservation of former eu- or supratidal morphological features (rock pools, rims, benches),
- Subsequent development of destroyed coastal

morphologic features (notches, benches, vermetid rims) and subsequent bioerosive sculpturing of the supratidal environment (rock pools),

- Washing out or removing of fine sediments (sand, shelly sand) within the deposits,

- Younger sediment deposition on top or in between the tsunami debris,

- Erosion of soil or vegetation destruction,

- Soil cover and soil development after deposition,

- Reworking of the deposits through subsequent storm or hurricane events,

- Dislocation of the deposits,

- Vegetation cover and stage of succession,

- Anthropogenic disturbance and relation to historic buildings and lighthouses,

- Historic reports (or lack of them).

4.5.2 Absolute Age Dating

In this first documentation about the tsunami deposits of the ABC-islands the determination of absolute ages can be regarded as pilot steps on the way to a detailed tsunami chronology. For a first overview about the advantages or disadvantages of dating coarse tsunami debris deposits, radiocarbon age determinations from 45 samples were performed from different geomorphologic units (boulders, ramparts, ridges) and on different material (vermetids, corals, gastropods).

Age determination was kindly performed by Dr. Bernd Kromer, Institut für Umweltphysik, Universität Heidelberg, Germany.

The age dating of coral debris incorporated in the deposits include some pitfalls, which has to be prevented through careful selection of the sample material. While these dates record ages of the material in the deposits, they do not necessarily represent depositional age(s). Without the possibility of stratigraphic control of sampling, it is difficult to decide whether the dated material is eroded and redeposited. In order to minimize the error margin the sample material has to show some qualitative characteristics:

- Preferably, samples from the branching coral *Acropora palmata* were taken, if these species did not occur in the deposits, the head coral *Diploria strigosa* was chosen. *Acropora palmata* samples are particularly suitable for radiocarbon dating because they consist of well-preserved original skeletons. Diagenesis or contamination is normally limited to the outer surface of the coral. It is easily recognized and can be removed before dating.

- The material was taken from the youngest growth zones, e.g. the tips of *Acropora palmata* branches, in order to define the age of breaking-off by the tsunami event as accurate as possible.

- The fine texture of the outer surface areas should be preserved and not altered by borings, encrustations and submarine cements, in order to exclude material, which has been deposited and eroded in the foreshore area over a longer time-period before the tsunami deposition.

More precise age data can be obtained from boulders, which show encrustations of *vermetids* and *bryozoa* colonies on the surface or consist entirely of these species. The incrustations indicate that the boulders were transported from the surf environment or from below sealevel. As soon as these organisms are not sufficiently supplied with seawater, they die and the dating of this material represents a reliable age data of the depositional event.

To determine the event as exact as possible it is important to take samples from well preserved outer parts of the vermetid or algal accretions, and therefore a careful cross-check with the different forms of living bio-constructions is necessary.

For the same reason, age dating of incorporated marine molluscs or gastropods without signs of bioerosion is preferable.

4.6 Microfossils

The incidental occurrence of deep-water microfossils can also be characteristic of paleotsunami deposits as shown by DAWSON et al. (1996) and DOMINEY-HOWES (1996). The intercalating sand facies within 13 tsunami deposits was examined for foraminifera by J.J. LOBENSTEIN, Amsterdam. Foraminifera species were present in small numbers, but no deep-water species could be identified (LOBENSTEIN, pers. comm.). Identifiable species include *Amphistegina lessonii* d'Orbigny and *Rotorbinella rosea*. The foraminifera valves were often broken, and therefore, identification was difficult. The destruction might be a sign of strong impact transportation, as supposed by DOMINEY-HOWES (1996) and others.

However, the lack of deepwater species, is not a prove against tsunami events, because the foreshore profiles along the ABC coastlines are very steep, and it seems nearly impossible to transport sediments from these subsea slopes upwards or even onshore.

5. Field Evidences

In this chapter the Holocene littoral coarse sediments and their depositional environment on the Leeward Lesser Antilles will be presented and documented in detail. Analog to their geographical position the islands will be described in a west-east order (Aruba, Curaçao, Bonaire), which corresponds with the magnitude of the paleotsunami events being the weakest in the west and the strongest in the east. Four different main types of debris formations have been observed within the debris deposits, each of these types has distinct geographic distributions, morphologic characteristics and internal compositions. Evidently the debris formations are the result of so-called "high magnitude-low frequency events", in this case most plausible tsunamis. As stated before, so far the debris formations have been exclusively attributed to hurricane-generated waves (DE BUISONJÉ, 1974). DE BUISONJÉ (1974), however, provides no supporting measured sections, maps or other field data to substantiate this hypothesis. Therefore, it has been necessary to establish a detailed documentation and characterization of the deposits based on intensive field studies accompanied by the application of dating routines (Chapter 6) in the present study.

First, a classification of the identified main debris formation types with a type illustration is given in order to correlate the descriptive characterization of the field evidences for each island according to the distinguished units in the summarizing subchapters. The debris formations will be described and evaluated with respect to their geographical distribution for each island separately (subdivided in the leeward and windward coast) from south to north in a clockwise direction, followed by a summarizing classification of the debris depositions and their distribution, which will be visualized and overviewed in form of distribution maps, boulder tables and graphics. Subsequently, a comparison of the islands will be given with regard to similarities and differences in:

- distribution and classification of debris deposition;

- transport width, inundation and runup heights as deduced from the debris distribution;

- energy values, origin and direction of the impact.

To accomplish this, the relation to other coastal forms together with associated distribution patterns will be documented. Furthermore, open questions concerning the debris deposits and the tsunami impacts are outlined. An overview of the location of the photographic documents (oblique aerial pictures, terrestrial photographs) is given for each island separately.

Characterization of Debris Deposits

1. Debris ridges with direct contact to coastline and surf zone (recent or subrecent)

1a. Recent
Asymmetric ridges of coral shingle with grain size distributions are ranging from centimeters to some decimeters with more gentle seaward slopes and steep inward faces. The inner edge is arcuate and in cases with some narrow tongues of shingle extending leeward over the fossil reef flat. Reworked ridges usually form much steeper sided structures. The orientation of the fragments is poor-sorted and they show no clear imbrication structures. These light-colored shingle ridges usually occur on the leeward sides of the islands with direct contact to the surf zone, where they are up to 15 m in width and 0.5-1.5 m high. The ridge itself may have coarse or cobble-sized regular and hemispherical shaped debris, but predominantly they consist of platy and stick-like coral rubble. The rubble originate from the living coral reefs, as e.g. observed after the youngest hurricane impact of *Lenny* in 1999, where large stands of living *Acropora cervicornis* were washed onshore.

However, not all material is necessarily growing at the time of the event, the bulk of the rubble is derived from debris on submarine terraces, ledges and in grooves. The debris is partly well-rounded as a result of the abrasion and show signs of bioerosion. A typical location is Pink Beach on Bonaire (Fig. 58).

1b. Subrecent
The typical subrecent debris ridges consist essentially of mostly well-rounded platy and rod-shaped coral fragments ranging from 10 cm to 60 cm in diameter with some rare limestone boulders present. The surfaces of the fragments are commonly weathered and blackened by the thin coat of endolithic filamentous algae. They occur along the southern and western leeward coastlines (in Bonaire also SE), situated directly subsequent to the shoreline, where they may extend over several hundred meters with width from 10-50 m and relative heights from 1-3 m. The mostly asymmetric ridges have convex or planar profiles with a steep seaward flank and a gentle dipping inner edge. The inner topography with heights of 0.5-1.0 m is chaotic and disordered. Frequently, an imbrication of the stacked fragments dipping to seaward can be observed and the orientation of the fragments is well defined. Predominantly, the rounded material is derived from debris deposits out of the

Fig. 58:
Debris ridge (nearly 1 m high) consisting chiefly of rods of *Acropora cervicornis* branches with tongues of shingle. Pink Beach, leeward coast, Bonaire.

Fig. 59:
Subrecent debris ridge at Willemstoren, leeward coast of Bonaire. The steep seaward margin locally reaches 80° and the gentle sloping inward face may show angles of less then 10°.

subtidal environment. Wave attacks can reach and rework the seaward flank of the generally inactive ridges. Today they often form a natural protection of the coast, and in cases they are used as a naturally heightened roadway like in the south of Bonaire. The debris ridge located at Willemstoren, Bonaire, illustrate the type of the subrecent debris ridge with direct contact to the coast (Fig. 59).

2. Debris ridges located several meters above sealevel and in distance to the coastline (subrecent)

Subrecent, inactive, weathered and blackish debris ridges located 10-30 m inland and up to +5 to 6 m asl with steep flanks on the leeward and the seaward edges and a usually sharp crest. They are narrower than the ridges of type 1 with less than 10 m width and a relative height of up to 2 m. The most dominant fragments are coarse and consist of blackened cobble to boulder-sized massive coral colonies of up to 2 m in diameter with an imbrication dipping seawards. The weight of a single coral fragment may reach 1 ton. Small to medium-sized limestone boulders are a frequent component of these ridges. Smaller fragments of corals, vermetids, and coralline algae can be deposited on the ridges through the agency of more frequent lower magnitude storm events.

Locally, major storm events can rework these generally stable landforms, which is documented by light colors of the rocky surface on the basis of the seaward margin or overturned fragments in the front part of the ridge itself. The ridge in Watamula, Curaçao, exemplary illustrates this type of debris ridge (Fig. 60). Its extension is rather limited on all islands.

Chapter 5.1: Field Evidences – Aruba

Fig. 60:
Subrecent debris ridge showing a steep seaward flank and a more gentle dipping leeward margin. The light colors on the seaward basis of the ridge show the reworking and leeward migration of the ridge mainly by more frequent lower magnitude events.

Fig. 61:
Rampart formation at Dos Boka, windward coast of Curaçao, located at +6 m asl and about 40 m distant from the cliff front.

Fig. 62:
Boulder arrangement at the windward coast of Bonaire, Spelonk Lighthouse. One of the last areas, which is not disturbed by mining activities and show the original distribution of limestone boulders deposited by a high magnitude event.

3. Ramparts (subrecent)

In the geomorphologic literature the term "rampart" ("boulder rampart") is usually used for scattered debris depositions of small to medium size with a thickness of some cm to several decimeters and a planar gently land inwards sloping profile, which are located in a distinct distance from the present shoreline or active cliff front. On the Leeward Lesser Antilles they occur along extensive stretches of the exposed windward northern and eastern coastlines. They are located with their seaward margin in distance of at least 40-50 m from the active shoreline, in cases up to 100 m, at elevations usually ranging from +6.0 to +10.0 m asl. These deposits extend the furthest land inwards of all observed depositional types, in Bonaire up to 200 m. The width of the up to 1 m thick deposits may exceed 40 m. Notably the blackened coarse material consist in contrast to the other deposition types predominantly of bad-preserved coral fragments and sharp-edged boulders derived from the supratidal zone with its rugged bioerosive surface. The size of the fragments reaches from 20 cm to 1 m with some rare larger boulders added. Frequently a sorting of the deposits according to the grain size with the largest fragments seawards and the smaller ones reaching further land inward can be observed. Imbrication and preferred orientation axis are not well defined. The state of weathering is significantly more intensive, most of the fragments are completely diagenetic altered from Aragonite to Calcite. The location of Dos Boka on the windward coast of Curaçao can considered to be typical for the rampart formations of the ABC-islands (Fig. 61).

4. Single boulders and boulder assemblages (subrecent)

Single boulders or boulder assemblages occurring all around the coastline in a zone reaching from the present shoreline up to more than 100 m inland at elevations of +2.0 to +12.0 m asl. The volume of the boulders ranges from 0.5 m^3 and can reach over 100 m^3. Exceptional large boulders may have a weight of nearly 300 tons. Most boulders have a box-size shape with a smooth base, smooth sides and a rugged phytokarsted upper surface. All boulders are formed exclusively of fossil limestone and contain diagenetic altered coral assemblages similar to those of the Lower Terrace, which forms the coastal platform on all islands.

The lithology indicates clearly that the boulders were derived from the coastal terrace. Morphologic features like rockpools and signs of bioerosion like the presence of sponge and *Lithophaga* borings and in cases encrustations of vermetids and coralline algae, give evidence that the boulders originated from the surf and spray-zone and the vertical cliff front of the coastal terrace and were moved onshore. The fact that the borings occur in most cases only on one or two sides suggests further that the boulders were broken out mechanically of the cliff front and were not derived form loose boulder material submerged in the shallow water in front of the cliff. Unfortunately, no *Lithophaga* shells were present in the borings that could be used for radiocarbon dating. Most boulders show signs of weathering as indicated by the darker surface colors. The boulder assemblages of the reference area Spelonk Lighthouse, windward coast of Bonaire, show impressively the huge amount of fragments deposited in distance to the present coastline (Fig. 62).

5.1 Aruba

Aruba is the smallest and most northwestern island of the Aruba-La Blanquilla island chain (see Fig. 1). The island, with its main axis in NW-SE direction, is part of the Venezuelan continental shelf. The distance between Aruba and the Venezuelan peninsula Paraguana is about 35 km and the maximal water depth in between is not more than 190 m. To the north and the east in a distance of only 10-15 km from the island the sea bottom drops gradually down from 1000 m to more than 4000 m in the Caribbean basin. Aruba has a length of 30 km and a width of up to 5.5 km. The island has a flat to slightly undulating morphology with low hills, "Jamanota" in the central part of the island is the highest hill and reaches 189 m (see Fig. 2).

Aruba is made up of a core of folded metamorphosed sedimentary and igneous rocks of Cretaceous age, unconformably overlain and surrounded by Eocene, Neogene and Quaternary limestone deposits (see Fig. 11). The Quaternary deposits form a threefold staircase of slightly uplifted Pleistocene coral reef terraces. Aruba's living Holocene coral reefs are rather restricted due to the location of the island on the continental shelf and to the high amount of sand in its shallow subtidal areas. The leeward reefs extend from Punta Basora to the south of Oranjestad and on the windward coast two small areas exist near California Lighthouse and at the south eastern tip of the island near Boka Grandi.

Aruba is almost continuously exposed to relatively strong trade winds, so all east facing shores experience medium to strong surf action with an average wave height of about 1.5 m and an average period of 7s (TERWINDT et al., 1984). The windward coast shows rocky and steeper morphology, where-

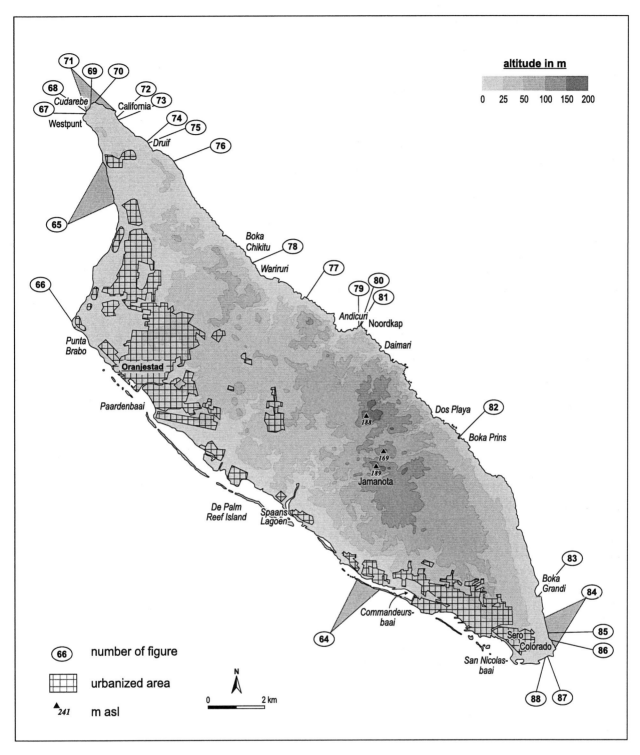

Fig. 63: Localization of all figures for Aruba.

as the leeward coast is dominated by a string of barrier islands and long stretches of calcareous sandy beaches, e.g. between Malmok and Oranjestad. The sandy beaches have been the basis for the development of mass tourism, which is the principle source of revenue; in total 95% of the leeward coastline has already been developed. Figure 63 overviews the locations of all photographic documents of this study for Aruba.

5.1.1 The Leeward Coast (S to N)

The Holocene debris deposits of the southwest coast of Aruba are restricted to a chain of small islands situated up to several 100 m off the coastline and extending from the village Sero Colorado in the south to the capital Oranjestad. These islands are 20 to 50 m wide, several km long and are mostly built up at the place of a submerged reef crest and

Fig. 64:
Small islands at the leeward side of Aruba showing older weathered ridges and some recent, light coral rubble from hurricane *Lenny* (1999).

consisting of coral debris deposits with some beachrocks present. Usually one debris ridge of weathered and blackish coral rubble is developed, but in cases a second ridge can be distinguished (Fig. 64). The origin of both ridges has to be placed into the Holocene. Recent debris deposits of predominantly *Acropora cervicornis* fragments can be found on the seaward side of the islands due to the impact of hurricane *Lenny* (1999), whereas at the landward sides mangrove vegetation is present, in particular in the shelter of De Palm Reef Island or near to Aruba airport. The mangrove belts seem to be undisturbed by sediment input or storm damage and showing aspects of an older well-established ecosystem.

Along the northwestern part of Aruba from south Oranjestad to the village of Malmok sandy beaches over long stretches have developed in connection with an extremely extensive submarine platform more than a kilometer wide and with a max. depth of only 20 m. Large areas of seagrass inhabit the sand sheets of the submarine platform (Fig. 65). Most of the touristic coastal infrastructure and hotel and apartment complexes are situated along this coastal stretch (Fig. 66).

In general, the leeward coastline of Aruba is low and without steep cliffs, only at the northern tip of the island near Arashi beach a cliff (about 1.5 to 3.0 m asl) with well developed notches in younger Pleistocene coral reef occurs. On top of the cliff in a distance of 6 m to the cliff front, a debris ridge with relative height of 1.5 m and a width of 20-40 m is located. Two different sediment units can be distinguished: The older one shows coarse coral fragments consisting of well-rounded massive corals. The size of the fragments usually not exceed more than 100 kg of weight, but some single boulders can reach weights up to several tons. All fragments are weathered and of dark to blackish color (Fig. 67).

At the seaward side of the ridge another type of sediments have been accumulated. Fresh looking light-colored coral fragments deposited at the seaward flank document that the ridge was reached by the wave activity during hurricane *Lenny* (1999) (Fig. 68). In front of the ridge scattered single boulders with a weight of up to 3 t were deposited.

Separated from the coastline by a small belt of vegetation, a rampart formation extending about 50-80 m inland is deposited on the Lower Terrace, which reaches a height of +5 m asl at this locality (Fig. 69). The rampart forms the equivalent of the coarse boulders in the ridge further south and is considered to be deposited during the same event. Around the northern tip the boulder rampart is transformed into a belt of more scattered fragments (Fig. 70). Where these boulders are numerous, the accompanying bench along the cliff shows signs of destruction.

5.1.2 The Windward Coast (N to S)

A characteristic landscape feature at the northernmost tip of Aruba near Hudishibana is a dune field – the California sand dunes – consisting of a multitude of steep and elongated dunes (Fig. 71). The material of the sand dunes originates from a narrow beach at the windward coast and is transported in an westerly direction by the constant trade winds; the sand of the beach is supplied by a living coral reef in short distance off the coastline. Close to the beach some huge diorite boulders are situated in the surf zone. The salt-resistant plant *Suriana maritima* plays an important role as a sand trap in the evolution of the sand dunes. The shrubby vegetation cover indicates

Fig. 65:
Shallow Holocene reef flat with extensive sand sheets in front of the beaches at the west coast of northern Aruba, mostly inhabited by seagrass fields. Living coral reefs are not abundant.

Fig. 66:
The broad spit of Manchebo beach at the west coast of Aruba with wide beaches built up by storm waves from coral sand and shingle of the extensive submarine platform.

Fig. 67:
South west of the California sand dunes in the north of Aruba fresh smaller coral fragments and weathered older and bigger ones are deposited on top of low cliffs.

Fig. 68:
The ridge of coral debris deposited by hurricane *Lenny* (November 1999) covering partly coarser and older coral boulders.

Fig. 69:
At the northernmost tip of Aruba an old weathered rampart of coral debris has partly been covered by vegetation. In the distance, the California sand dunes can be seen.

Fig. 70:
Oblique aerial photograph of a strip of boulders and partly destructed benches at the northern tip of Aruba island.

that these dunes are present since decades or centuries. The dunes partly cover a broad rampart of dark boulders, extending from the northernmost point of the island in some distance from the shoreline to the south east, as can be seen on the oblique aerial photograph of Fig. 71.

East of the California dune fields an elongated rampart formation is present in a distance of about 50 m from the cliff front (height +5.0 m asl), containing rounded coral debris and – in this being unique for all debris deposits of the ABC-islands – predominantly large amounts of bigger vermetid accretions originating from the bench (Fig. 72).

Along this coastal stretch greater parts of the bench are completely destroyed. The rampart reaches up to 200 m inland with a sorting of the bigger boulders (~10 t) deposited closer to the shoreline and smaller ones further inland (Fig. 73). At two of the boulders vermetid accretions have been dated by radiocarbon dating (see Chapter 6). All fragments and boulders are strongly weathered and the dark surfaces show signs of an intensive karstification process.

Further south (Location Druif) the rampart formation is replaced by a debris ridge with a maximum relative elevation of more than 2.0 m, a width of up to 50 m and in places rather steep slopes (Fig. 74). The ridge, deposited 30-40 m from the coastline on top of the Lower Terrace (height +3.0 m asl), consists of coral rubble of different sizes ranging from several kg up to about 1-2 tons. The rubble show rounded fragments as well as fine sculptured coral pieces and in smaller amount parts of vermetid rims from the bench in front of the cliff (Fig. 75). Occasionally limestone boulders of 1-2 t can be found in the ridge. The seaward slope show lighter colors due to the deposition of finer material including shell sand (Fig. 76). The boulder ridge extends further south east at the foot of the gabbro hills for some kilometers, sometimes transformed into a flat rampart. Large boulders are common, the largest one found on Aruba with a weight of slightly more than 15 tons lies near Boka Pos di Noord.

At this place some general conclusions about the field observations will be discussed as these findings play a key role in our argumentation concerning the depositional process and the age of the debris deposits. Evidently all larger fragments have not moved for a longer time period as indicated by the very dark color caused by a layer of endolithic algae. The surface of the coral fragments and the limestone boulders is in most cases heavily weathered and the fine sculpturing of the coral species is hardly recognizable. The surface of the Pleistocene Lower Terrace (isotope stage 5) forms the basis of the debris deposits. Almost everywhere a young karst surface, caused by subaerial solution through rainwater, with small peaks of limestone protruding not more than about one decimeter from the depressions lying in between has developed. Within the splash and spray zone of the supratidal another karst phenomenon can be distinguished, which is characterized by sharp, irregular limestone peaks with rockpool depressions in between. The rough microtopography is the result of bioerosion and often reaches 60-80 cm, in cases up to 1.0 m. Here, fossil corals such as *Acropora palmata* might be dressed out of the limestone "en relief" and often form the top of the limestone peaks. Between the debris deposits in form of the ridges with their well-defined extend (often situated in a distance of 20-40 m from the coastline) and the cliff, this rough sculptured surface zone is exposed, but strikingly completely free of sediments, although the rockpool depression would represent an excellent sediment trap for coarse material.

It can be concluded that waves – either normal or storm-induced – were not capable to transport and deposit coarser material on top of the Lower Terrace from the foreshore zone and no fragments have been broken off from the cliff profile during the recent history. Another suggestion would be that no rubble sediments are present in the foreshore zone of all three islands. The later assumption is highly unlikely due to living coral reef growth even though in lesser extension. These sediment free areas accompanying the windward cliff coasts are a significant feature of all three islands. That does not exclude that singularly boulders or coral rubble of smaller sizes may occur from now and then, but nowhere accumulations of larger amounts have been observed.

Along the coastline further south up to the height of the Natural Bridge area, ramparts are the dominant depositional type. A scattered line of bigger boulders near the sea front sometimes accompanies the ramparts. Obviously the most significant accumulation of boulders are corresponding to stretches of the coastline with bench destructions or cliff collapses. With undulations of the cliff front the rampart front or the row of boulders bends back and traces the contour of the coastline in a certain distance not exactly parallel, but slightly shifted sideward. From this observation the conclusion can be drawn that the forces, which destroyed the bench or the cliff front and transported the boulders upwards, did not approach perpendicular to the coastline, but from a more easterly direction (Fig. 77). However, at many places the reconstruction of the deposits of coarse rubble and boulders is difficult because of the extensive mining activities (Fig. 78).

Fig. 71:
Oblique aerial photograph of northern Aruba (view to southeast). The strictly east-west-shifting California sand dunes partly cover a broad and old boulder rampart seen as a belt of dark gray colors.

Fig. 72:
East of the California sand dunes, a scattered field of boulders is deposited up to 200 m from the cliff front.

Fig. 73:
Huge boulders from a low cliff, up to 10 t in weight. Sand deposits in the shelter of the boulders generate the light colors in the area. In the distance the California lighthouse.

Fig. 74:
Landward slope of the weathered boulder ridge near Druif, windward coast of northern Aruba. Grass vegetation is settling on the lower parts of the landward flank of the ridge.

Fig. 75:
The old boulder ridge south of Druif contains rounded material as well as well preserved corals (*Diploria* sp.) and fragments of vermetid benches (with hammer).

Fig. 76:
At the seaward slope of the older boulder ridge bigger fragments as well as sand deposits originating from the lowermost strata are situated. South of Druif.

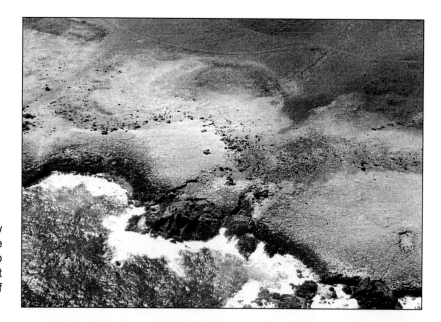

Fig. 77:
Lines of boulders or ramparts follow the contours of the cliff line, but have been slightly shifted aside due to wave impact approaching from left (the east). Central windward coast of Aruba. Oblique aerial photograph.

Fig. 78:
Many of the rampart formations along the windward side of Aruba have been removed by mining. Oblique aerial photograph.

Fig. 79:
Oblique aerial photograph of the east part of Andicuri Bay, south of Natural Bridge, central east coast of Aruba. Boulders on top of a Pleistocene coral reef flat lie at +9 m asl. The reddish colors show remnants of a fossil soil covered with boulders and rampart debris.

Along the windward coast, in particular in the outcrop areas of the gabbro and diabase-schist-tuff formations, only a narrow ledge of Lower Terrace limestone is preserved. In the area of Andicuri Bay in the central part of the windward coast the discordance between the youngest Pleistocene coral reef terrace and the non-carbonate basement is exposed. Along this level percolating fresh water excavated shallow gullies and in some cases these gullies drain underneath the limestone of the Lower Terrace. As the unconformity is exposed a little bit above sealevel, waves are simultaneously eroding along this plane and have developed several natural bridges of different sizes. This phenomenon first has been described by MARTIN (1888).

Within Andicuri Bay some single boulders as well as remnants of ramparts can be seen, here the Lower Terrace has a height of +7 to +8 m asl. At the eastern part of the bay close to Noordkaap a debris ridge together with some larger boulders up to some tons are deposited at a maximum height of +9 m (Fig. 79). Landinward reddish soils appear subsequent to the gray fossil reef crest. These deposits are relicts of a former deep loamy soil formation of a terra rossa type, which developed behind the protective position of the reef crest. During the deposition of the debris ridge the soil remnants have not been eroded, which is only possible if the deposition has taken place in a very short time period or only very few waves were responsible for the deposits. In contrast, the soil relicts are completely washed out, where presently reached by the surf, but no material like well rounded pieces of diabas (Fig. 80) or sand and pebbles (Fig. 81) are deposited up to this height or on the soil relicts under present conditions.

Within the older debris ridge many gastropods, in particular the species *Cittarium pica*, have been incorporated into the sediment. A sample was taken for radiocarbon dating (see Chapter 6). These gastropods inhabit the rocky foreshore environment.

Southwards from Noordkaap the coastline is characterized by outcrops of diabas rocks, which are shaped in typical "woolsack" forms by deep chemical weathering and are now exposed due to subsequent denudation. North west of Dos Playa the Lower coral reef terrace appears again and become broader to the southeast. From here to Boka Prins rampart deposits are developed at many places, typically in some distance to the cliff front and several meters above the present sealevel and reaching inland up to 100 m (Fig. 82).

Further southward from Boka Prins to the south east corner of Aruba island all three Pleistocene Terrace complexes come into sight again, surrounding and attaching the 47 m high Sero Colorado rock formation, composed of deep lateritic weathered diabas and gabbro. The multi-colored surface (predominantly reddish) has given the cape its name. Between Rincon and Boka Grandi extended beaches (partly with beachrocks and drifting sands) have been developed supplied by a living coral reef in a short distance offshore (Fig. 83). South of Boka Grandi again rampart formations occur, but the size of the scattered boulders remains rather small (~0.5 t), unless the accumulations are cut off due to the protective position of Sero Colorado. One fragment of 7.7 t broke during the depositional process into 4 pieces; attached parts of vermetid accretions were sampled for radiocarbon dating (see Chapter 6). The boulder was moved 70 m land inwards up to a height of +4 m asl. During the transport the block was overturned, which is documented by rockpool features now situated on the bottom of the boulder. The surface shows signs of fine-sculptured karst weathering of 2 cm depth.

Along the last kilometer north of Sero Colorado the lower coral reef terrace shows small steps and a straight cliff line accompanied by a beautiful example of a horizontal bench with a width of up to 12 m (Figs. 84 to 86). The presence of the bench gives evidence for a stable relative sealevel still stand over a long time period. The bench is most perfectly developed to the south, where the water depth in front of the cliff increases and abrasive material that is moved within the surf zone becomes less abundant.

South of Sero Colorado a horizontal destructive bench has been developed in the non-carbonate rock formations. The bench reflects the level of permanent saturation in the rock and therefore the level of salt weathering reaching down to this benchmark. The flat surface of the bench lies about 1.5 m asl and is moistened constantly up to this level by the trade wind surf.

At the western side of Sero Colorado two parallel and steep ridges consisting of a chaotic mixture of coral shingle, rubble and large fragments of *Acropora palmata* have been deposited (Figs. 87, 88). The elongated ridges have a maximum relative height of nearly 3.0 m and are covered and consolidated by vegetation, documenting a deposition at least several decades ago, nevertheless an artificial origin can't be entirely excluded, but seems to be highly unlikely.

We suggest a deposition during single extreme events, which have no similar counterparts during the recent history – most plausible a tsunami. Further

Fig. 80:
Top of the Andicuri rampart and ridge with boulders up to 6 tons at 9 m asl. The black rounded fragments are diabas boulders derived from the foreshore environment.

Fig. 81:
Detailed look at the ridge deposits of Andicuri Bay at 9 m asl: well-rounded coral and diabas pebbles and cobbles, a coral head of *Diploria strigosa*, and many broken *Cittarium pica* shells.

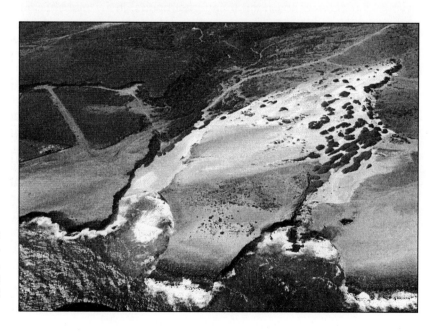

Fig. 82:
Aerial view of Boka Prins: windblown sand sheets and remnants of a boulder rampart between the two bays. See the asymmetrical shape of the bays due to wave impact from left (east).

Chapter 5.1: Field Evidences – Aruba

Fig. 83:
Boka Grandi with beaches and vegetated dunes in front of a fossil cliff in a Pleistocene coral reef terrace. The surf aspect shows shallow water conditions behind a fringing Holocene coral reef.

Fig. 84:
Sero Colorado (left), the most southeastern point of Aruba island with the attached lower coral reef terrace to the east. Note the perfectly developed bench. The gray colors of the old reef flat show scattered rampart deposits.

Fig. 85:
Close-up oblique aerial view of the bench and ramparts north of Sero Colorado.

Fig. 86:
Terrestrial view along the bench north of Sero Colorado. Scattered smaller boulders on the left in distance to the cliff at around +5 to +7 m asl.

Fig. 87:
Double ridges west of Sero Colorado in front of the lowest Pleistocene coral reef terrace.

Fig. 88:
Coral sand and weathered *Acropora* boulders building the double ridge west of Sero Colorado. The vegetation shows that the deposition has been taken place at least many decades ago.

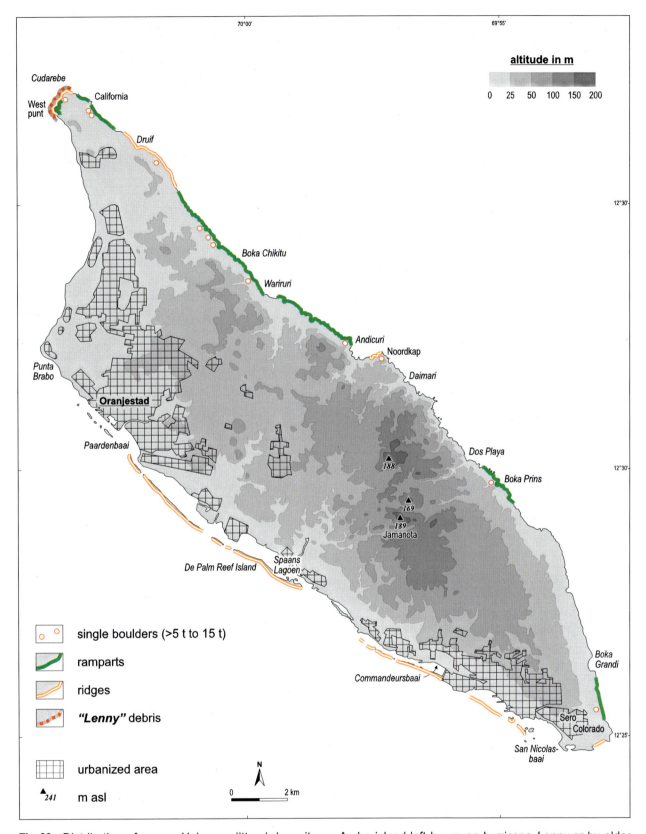

Fig. 89: Distribution of coarse Holocene littoral deposits on Aruba island left by young hurricane *Lenny* or by older tsunamis.

investigations of the sedimentary record of the ridges might highlight which depositional process the deposits owe their existence. Immediately to the west of these ridges at the location of a pet cemetery another chaotic deposition of sand with rubble and boulders with no distinctive morphology exists, now covered by vegetation. Its origin remains to be investigated as well.

Tab. 12: List of large tsunami boulders from different sites along the shoreline of Aruba.

Location	Nr.	a [cm]	b [cm]	c [cm]	volume [m³]	weight [t]	height [m asl]	distance to coastline [m]
Arasji	1	200	160	50	1.6	4.0	4.5	20
	2	270	180	40	1.9	4.9	4.5	25
California Lighthouse	3	470	140	60	3.9	9.9	5	80
	4	280	160	80	3.6	9.0	4.5	75
	5	300	130	120	4.7	11.7	4	20
	5a	260	160	120	5.0	12.5	4	20
Boka Poos di Noord	6	320	160	120	6.1	15.4	2.5	25
Natural Bridge	7	300	180	50	2.7	6.8	7	32
Sero Colorado	8	410	150	50	3.1	7.7	4.5	60

5.1.3 Summary and Conclusion

On Aruba coarse Holocene littoral debris deposits are mainly restricted to the windward coast, where they are developed as subrecent rampart formations or as subrecent debris ridges located several meters above sealevel and in some distance to the coastline (Fig. 89). At the bending to the leeward coast of the northernmost and southernmost island tips and along some narrow islands at the leeward side, as well ridges of this type have been deposited. Young coarse sediments like fresh coral debris only occur in a narrow zone close to the present waterline or on top of low cliffs. The fresh material with its light colors has been supplied and deposited onshore recently by hurricane *Lenny* in November 1999, whereas the process of deposition and the age of the subrecent deposits is the object of the discussion in this study.

The hydrodynamic estimation of the wave heights for boulder movement of the boulder sizes occurring on Aruba clearly indicates that storm-induced waves with wave heights between 13-56 m would have been required to just overturn the boulders (see Tab. 12). Either extreme hurricane events should be considered, but these intensity scales as implied by the wave heights are until present never observed worldwide or the hurricane intensity in the Caribbean region must have been significantly higher in the past, for which no evidence was found.

Generally, hurricane events are not very common on the ABC-islands. They occur in time distances of decades and even then their track is usually in a certain distance to the islands themselves – in known history no landfall of a hurricane was described. The location of the deposits several decameters in a distance to the recent coastline and their strictly separation from the modern surf regime might suggest a former higher sealevel, but all field evidences (benches, notches) refer to a stable long lasting sealevel still stand.

Therefore, we state that tsunamigenic origin should be considered as the most convincing source for the debris deposits, as tsunami events are extremely rare and highly energetic compared to hurricanes. These events are well able to break off cliffs and benches and transport boulders far inland as evidently occurred in Aruba. Radiocarbon dating will help to clarify how many events are represented within the coarse and weathered deposits.

5.2 Curaçao

Curaçao, the largest of the three Leeward islands of the Netherlands Antilles, is situated between Aruba (still part of the Venezuelan shelf) to the west and Bonaire, from which the island is separated by the Bonaire Basin with a depth of 2000 m, to the east (see Fig. 8). The subsea slopes of the island are very steep (Fig. 90) and occur in -900 to -60 m depth as 10°-20° mud and sand slopes or 60°-90° rock slopes (REED, 2001). In shallow water from intertidal to depths of ~50 m the deep fore reef generally has a 45°-90° slope and at several sides the steep slope continues below the 50 m depth contour.

Curaçao extends from SE to NW for about 61 km and the width is varying between 4 km in the center and about 11 km at both the NE and SW island parts. Two large anticlines of the Curaçao Lava Formation form these parts of the island, which were formerly separated and connected during Younger Pleistocene times by ongoing uplift in a number of steps and the development of fringing coral reefs. These fossil and elevated coral reefs now form a terrace sequence from the Highest Terrace (150-90 m) over the Higher Terrace (80-50 m) and Middle Terrace (25 m) to the Lower Terrace (10 m) (see Fig. 13). Recently the age of the Lower Terrace was determined by ESR-dating to isotope stage 7 at the base and isotope stage 5e for the upper parts of the reef deposits (SCHELLMANN, RADTKE & SCHEFFERS, in press). Living coral reefs are developed as steep sloped fringing reefs and situated close to the shore (generally within a distance of 100 m) along the leeward coast up to a depth of about 60 m. Along the northeast coast coral reef growth is rather limited, and instead calcareous sediments are covered with large fields of the reefal alga *Sargassum* down to depths of -10 m (BAK, 1975).

The coastline of Curaçao generally is formed by carbonate rocks of the Lower Terrace and predominantly rocky and steep. A deeply incised notch can be found nearly all around the coastline in the elevated Lower Terrace limestones. Along the trade wind exposed north- and east facing shores benches usually occur. Below the terraced bench a second notch is present as a result of abrasive and bioerosive action. The Lower Terrace limestones are dissected more or less perpendicular to the coastline by short and steep-sided bokas (mostly in the northern windward part of the island). Larger boka systems are developed almost exclusively as lobate, hand-shaped inland bays (or ria-like features), e.g. the Sint Jorisbaai on the windward coast. These forms, however, are also common to a smaller extend in the northern parts on the leeward side, but more branched and reaching further inland, e.g. Santa Marta Baai, Piscaderabaai, Schottegat, the smaller Jan Thiel Baai, or the extended Spaanse Water. At the southern tip of the island three elongated, coast-parallel lagoons, Fuikbaai, Awa Blanku and Awa di Oostpunt are formed. Furthermore, two semicircle bays characterize the leeward coast: The larger one is Bullenbaai, evi-

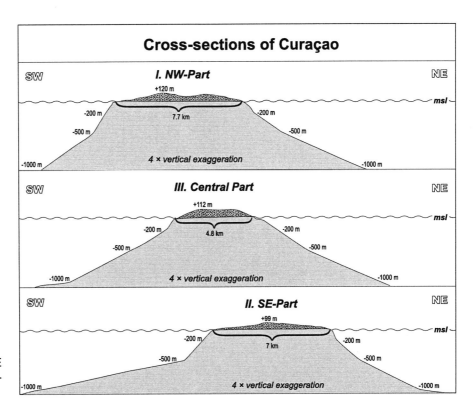

Fig. 90:
Three cross-sections (NW, SE and Central Part) over the island of Curaçao.

dently an old feature, because the bay is surrounded by at least two Pleistocene reef terraces. Caracasbaai, the smaller one, located directly west of Spaanse Water doesn't possess fossil reef terraces. The origin of this bay will be discussed in Chapter 5.2.1.1.

Mangrove habitats occur in the shore zone of the inland bays, but their extend often is limited caused by infrastructure activities. By far the most common mangrove species is the red mangrove (*Rhizophora mangle*).

Due to the paucity of natural sandy beaches on Curaçao, a trend has developed since 1986 to create artificial beaches for tourism using containment dams in areas where beaches were not formerly present (DEBROT & SYSBEMA, 2000). The locations of all figures are positioned in Fig. 91.

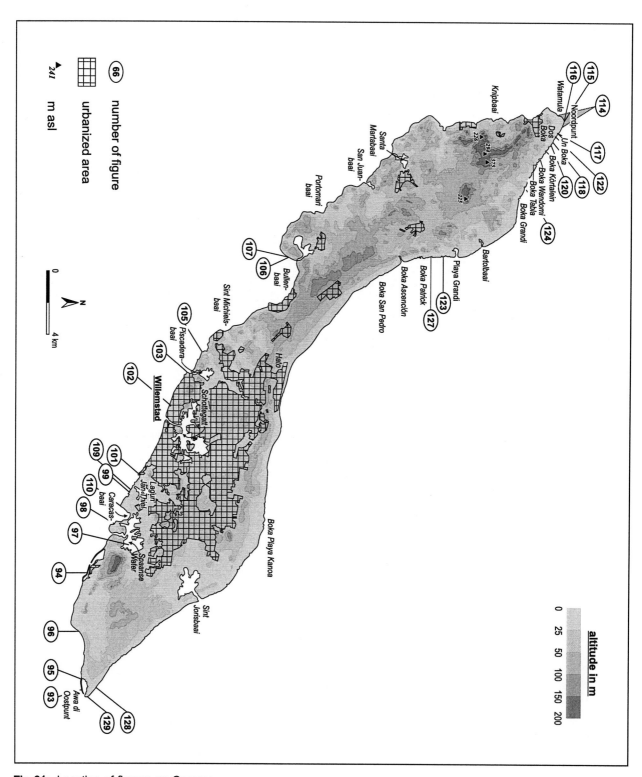

Fig. 91: Location of figures on Curaçao.

5.2.1 The Leeward Coast (S to N)

Oostpunt to Lagun Jan Thiel

Along the coastal stretch from Oostpunt to Lagun Jan Thiel the most extensive ridge deposits of the entire leeward coastlines of the ABC-islands are situated, only on the southern side of Klein-Bonaire comparable deposits can be found (Fig. 92).

The inner bays Fuikbaai and Awa di Oostpunt and the lagoon of Awa Blanku are delineating the coastline of this area. In contrast to Aruba, where a line of narrow, elongated islands accompanies the coastline, these ridge-like features along Curaçao are adapted to the mainland either in form of a damming lagoon (Awa Blanku) or spit-like with a small opening to the sea (Awa di Oostpunt, Fuikbaai) (Fig. 93). The basis of the barriers and spits is built by the youngest Pleistocene coral reef (<2.0 m asl) and mostly covered by recent and subrecent coral rubble.

The recent rubble deposits are the result of younger storm activity, which accumulated predominantly *Acropora cervicornis* fragments onshore in ridges with a maximum height of 1.0 m and steep profile at both sides. At some places, smaller boulders of the Pleistocene reef terrace are incorporated, but they have not been moved during the recent events as evident by the karst microtopography of the surface. In contrast, the subrecent ridges are remarkable wider with a max. height of 3.0 m and a width of 40-60 m. The material is like-wise predominantly composed of *Acropora cervicornis* and in lower amounts *Acropora palmata* and *Diploria*-species can be found.

At location Fuikbaai, the percentage of coarser fragments increases towards the eastern beginning of the barrier. Further on, limestone boulders up to 8 t may be incorporated. For the last several hundred meters, the deposits are completely disturbed by digging activities for military purposes during World War II, since then they were not removed. In fresh vertical sections the penetration depth of the weathering front is visible, reaching only several centimeters into the debris (Fig. 94).

On subrecent tsunami ridges in Bonaire, we observed in contrast a penetration depth reaching some decimeters deep into the debris fragments. At the western end of Awa di Oostpunt spit the rubble ridge extends, now up to 100 m wide, situated on the fossil coral reef platform at a height of 3.0-3.5 m asl (Fig. 95). Two samples for radiocarbon dating were taken at this location (see Chapter 6).

In direction Fuikbaai, the ridge divides at places into two or three parallel ridge deposits, separated by a narrow line of shrubby vegetation or deposits of finer sediment covered with grass (Fig. 96).

In particular on the stretch between Awa di Oostpunt and Awa Blanku the double ridge character is clearly visible. Evidently, the landward ridge with a relative elevation of +3.0 m is of older age as indicated by the denser shrub vegetation. The most exposed flanks of the seaward ridge are reached and partly reworked by recent storm surf action. Only sometimes, a sediment free section of fossil coral rock is exposed along a cliff of +0.5 m height, from which locally boulders have been broken off and deposited further inland. The rubble deposits extend to the entrance of the inland bay Spaanse Water (Fig. 97) and from there further until the eastern part of Caracasbaai (Fig. 98).

At the western side of Caracasbaai close to the area of lighthouse Lyhoek, several boulders (2 to 10 t) are situated in a distance of 60 m from the coastline at a height of +6.0 m. Another subrecent ridge starts 500 m further following the coastline in direction Lagun Jan Thiel at a height of +4.0 m, reaching a relative height of 1.5 m to the crest (Fig. 99). The material is coarse with boulders of *Acropora palmata* of up to 1 ton, and occasionally larger boulders (~6 t) are deposited in a distance of 50-60 m from the coastline hidden in the accompanying shrub vegetation. The highest deposits of recent storm activities are reaching the basis of the ridge only occasionally. At Lagun Jan Thiel the deposition of subrecent rubble ridges ends abruptly.

In the middle part of the tsunami ridge in a distance of 30 m from the cliff front at a height of +4.5 m several hundred fragments (10-20 cm length) have been measured in order to determine the source of the material. The material was distinguished into seven classes:

1. Very well preserved coral species;

2. Bad preserved coral species;

3. Well-rounded fragments;

4. Bioerosive sculpturing predominant;

5. Shape heavily weathered by karstifikation;

6. Debris fragmented at one side;

7. Debris fragmented at several sides.

Figure 100 illustrates the percentage of the different types of the material: In total, 677 fragments were counted. 8% show a perfect sculpturing, 28% are bad preserved, but still identifiable (predominantly *Acropora cervicornis*). Only 4% are well rounded, but 19% show signs of bioerosion. 3.5% are broken at one side, whereas 28.5% are broken at several sides. Significant influence of karstification could be found at 9% of all fragments. The results clarify that at least

91% of the material has been transported out of the subtidal environment and is of Holocene age. Only maximal 9% are broken out of the Pleistocene reef terrace, which has a height of 2.0 m at this location.

The living coral reef along the coastal stretch from Oostpunt to Lagun Jan Thiel is highly biodiversive as a result of the exposed situation and characterized by the dominance of *Acropora palmata* communities.

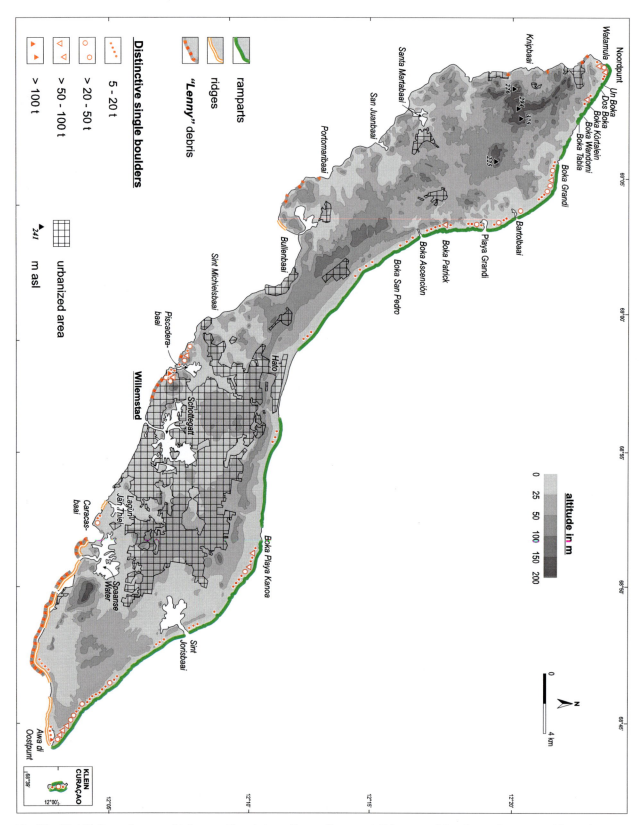

Fig. 92: Distribution of coarse Holocene littoral deposits on Curaçao left by recent hurricane *Lenny* or older subrecent tsunamis.

Chapter 5.2: Field Evidences – Curaçao

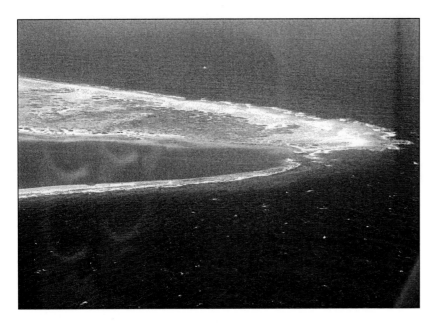

Fig. 93:
Aerial view of Awa di Oostpunt with the spit-like ridge enclosing the inner bay area.

Fig. 94:
The weathering front penetrates the debris deposits at Fuikbaai only for some centimeters.

Fig. 95:
Rubble ridge to the west of Awa di Oostpunt showing a steep crest and a convex profile.

Fig. 96:
Oblique aerial picture of the double ridges between Awa di Oostpunt and Awa Blanku. The arrows mark the two distinguished deposits, which are separated by a narrow line of vegetation.

Fig. 97:
Oblique aerial view of Spaanse water, a ria-like inland bay, which is intensively developed for housing projects and a private harbor.

Fig. 98:
Oblique aerial photograph of eastern Caracasbaai, with the subrecent ridge deposits on top of the Lower Terrace at a height of +3.5 m asl.

Fig. 99:
The subrecent ridge deposits at Lagun Jan Thiel consisting chiefly of rods of *Acropora* branches.

The reef flat is rather narrow (~25 m) with the drop off occurring in 8-10 m depth and adjacent slope angles of 20-40°. A shore zone community, which comprises composite patches with sand, coral debris, beach rock and hard bottoms dominates the upper reef flat (VAN DUYL, 1985). Scattered corals such as *Diploria strigosa, Diploria clivosa, Siderastrea radians* and *Porites astreoides* are common. This community occurs in all wave energy environments along slowly rising coasts, whereas it is absent adjacent to cliff coasts, where the water depth directly exceeds 2 m (VAN DUYL, 1985).

Two factors might be responsible for the better development of the subrecent ridge deposits on this leeward coastal section on Curaçao compared to Aruba and Bonaire: The exposed oceanographic conditions allow the growth of a flourishing coral reef with a high cover percentage of mainly massive and branching corals (*Acropora palmata*), whereas on Aruba the leeward reef development is restricted to patchy reefs of minor extent. On Bonaire, due to the N-S orientation of the island, the leeward coast show rather calm wave energy environments with coral reef structures dominated by less massive *Acropora cervicornis* communities.

Secondly, a higher energy impact of a tsunami wave attack because of the exposed conditions might be responsible for an increase in the ability of debris transportation in this coastal section.

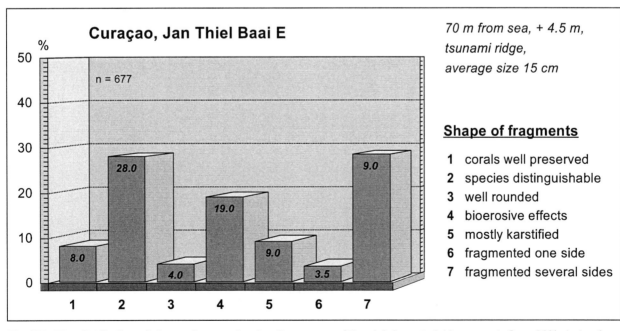

Fig. 100: The distribution of shape classes showing the source of the debris material in percent. Over 90% derive from the Holocene reef environment.

Willemstad and surrounding areas

Opposite the Sea Aquarium south of Willemstad the only evidence for a Holocene sealevel dislocation on the Leeward Lesser Antilles is documented with a small notch uplifted for about 0.4 m over a short distance in the cliff front (Fig. 101). In general, the coastal zone around Willemstad is entirely transformed by building activities and natural coastal sections rarely occur. A small *Lenny*-deposition appears along a barrier and lagoon with mangrove shrubs directly west of Willemstad close to the water factory (Fig. 102).

Piscaderabaai to Slangenbaai

Boulder assemblages of remarkable size considering the leeward exposure are located along this coastal section. The boulders are deposited on the Lower Terrace unit, which has a height of +2 m and is undercut by a deep notch with a sheltered profile. The age of the Lower Terrace deposits is dated with the ESR-dating routine to isotope stage 5e (116 ± 11 ka, SCHELLMANN, RADTKE & SCHEFFERS, in press). Several huge boulders of Pleistocene coral reef rock can be found on the domain of the Sheraton Hotel at the eastern side of Piscaderabaai. With a weight of 194 t and a longest axis of 6.5 m, here one of the largest boulders of Curaçao is deposited close to the cliff and presently used as a lookout (Fig. 103).

Other boulders with a weight of 84, 44 and 23 tons respectively, are situated 50 m inland and constructional incorporated in the walls of a ruin, which belong to the remnants of Fort Piscadera, built more than 300 years ago (Fig. 104).

Fig. 101:
Small notch opposite the Sea Aquarium indicating a Holocene sealevel dislocation of 0.4 m. This location shows the only evidence of a Holocene dislocation we observed on the Leeward Lesser Antilles.

Fig. 102:
Lenny deposits on the coastal section behind the water factory. The barrier ridge is colonized by mangrove species.

The deposition of boulders of considerable size (8 - 67 t) can be followed west of Piscaderabaai until Slangenbaai, where a cluster of boulders is deposited in the continuation of a small incision in the coastline, with the largest boulder reaching a weight of 18 tons (Fig. 105). These boulder assemblages mark the end of subrecent boulder deposits along the entire leeward coast.

Bullenbaai to Watamula

At the western side of Bullenbaai, a half-moon shaped bay, the last subrecent ridge deposits of the leeward coast appear. Depositional evidence of hurricane *Lenny* is absent on this coastal section due to the *Lenny*-sheltered south-eastern exposure of the coastline. The subrecent ridge deposits are 30-40 m wide on average, but extremely disturbed by mining activities (Fig. 106).

Fig. 103:
Largest boulder perched on the leeward coast of Curaçao at the domain of the Sheraton Hotel. This boulder is 6.5 × 4.6 × 2.6 m in size with a weight of 194 tons.

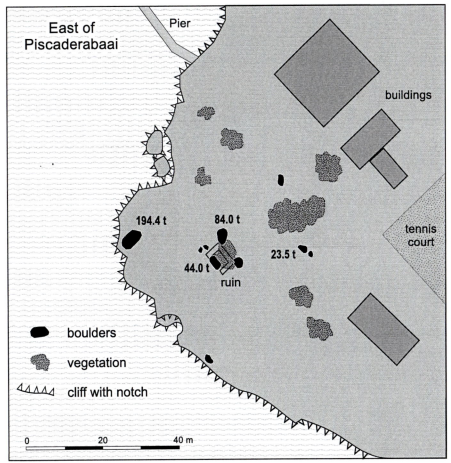

Fig. 104:
Boulder assemblage around the remnants of a small ruin belonging to Fort Piscadera, which was built in the 17th century.

Fig. 105:
Boulders resting on the Lower Terrace deposits at a height of +3.0 m asl at Slangenbaai.

Fig. 106:
At the western side of Bullenbaai extensive rubble ridges have been deposited and are now colonized by shrub vegetation indicating a subrecent age of the rubble.

Fig. 107:
Debris accumulations in front of a paleocliff at Bullenbaai. The deposits show imbrication and have an upward sloping profile in direction of the paleocliff.

In contrast to all other tsunami ridges, they are deposited against a steep fossil cliff, sometimes with a steep upward sloping profile. The highest undisturbed parts reach 3.0 m asl. Some single boulders (~1 ton) and large *Acropora palmata* fragments can be situated on top of the fossil cliff at a height of +6.0 m. The debris composition shows a differentiation between the seaward and the landward sides of the ridge: On the seaward side *Acropora cervicornis* branches dominate (~90%), whereas landward much coarser fragments of *Acropora palmata* cover 60-70% of the surface area.

Shape measurements have been conducted to determine from which environment the debris derives. The first measurement was carried out at a height of +2.5 m asl in a distance of 30 m from the coastline.

From 653 fragments with an average size of 30 cm (max. 50 cm) 2% show a well-preserved surface (mostly *Diploria* sp.), 26% are bad preserved, but still identifiable (*Acropora palmata* and *Acropora cervicornis*), only 1% are well rounded, whereas 18% show bioerosive destruction and 9% show significant karstification (Fig. 108).

The damage estimation overviews that 11% are broken at one side, and 33% of the fragments at several sides. The result suggests that at least 90% of the debris derives from the near fringing reef and the foreshore environment and is therefore of Holocene age. A second area was analyzed further inland in 50 m distance to the coastline at a height of 2.9 m. The size of the particles increased slightly with an

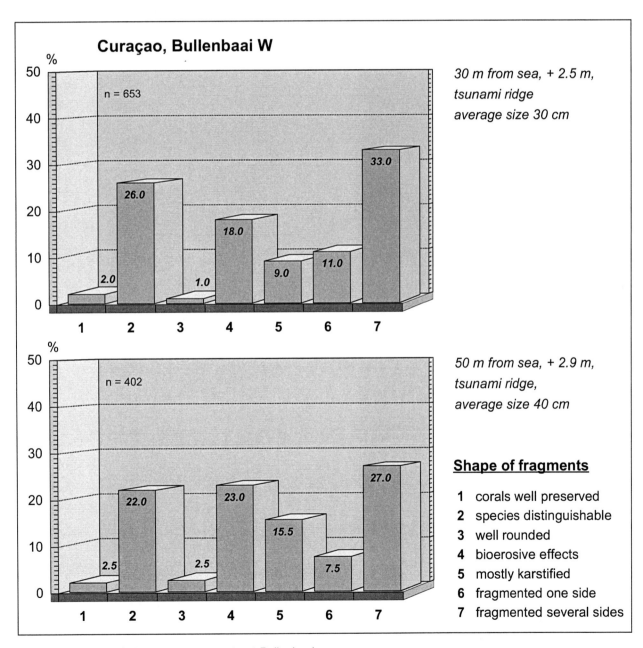

Fig. 108: Diagram of shape measurements at Bullenbaai.

average of 40 cm to max. >60 cm. This part of the ridge shows a slope of 10% dipping seaward. Due to the coarser size on the same surface area of 16 m², only 402 fragments have been counted: Very well preserved are 2.5% and 22% are still distinguishable. Class 3 with well-rounded shapes contains 2.5%, 23% in class 4 show bioerosive altering. An intensively weathered karst surface could be observed at 15.5%. Broken at one side are 7.5% compared to 27% damaged several sides. It can be concluded that more than 85% of all fragments derive from the Holocene fringing reef. The results confirm the measurements at Jan Thiel Baai.

From Bullenbaai the lowermost fossil reef platform rises in height to +6.0 m asl, now with a steep vertical cliff front developed. Deep bioerosive horizontal notches (sheltered profile) undercut the cliff and are likewise developed in boulders deposited in the subtidal environment, giving them the shape of mushroom rocks. On the top of the terrace, no subrecent or recent deposits are accumulated. Only along the beaches of the smaller bays (e.g. Portomariebaai, Playa Grote Knip, Kleine Knip, Playa Kalki) fresh hurricane deposits are found, resulting predominantly from the destruction of the reef flat coral communities during hurricane *Lenny* in 1999 (Fig. 109).

As most beaches are managed for recreational purposes, the *Lenny* deposits have been removed or transformed and the original depositional extent is difficult to reconstruct. At Watamula the leeward exposure bends to the exposed windward island side and the extent and size of the debris deposits will increase significantly with the strongest impact on the northern and southern tip of the island.

Fig. 109:
Ridge of fresh coral rubble (mostly *Acropora cervicornis*) from Hurricane *Lenny* (Nov. 1999) with a maximum height of 0.6 m above high water level, partly affecting living vegetation. The picture was taken 90 days after the event.

Fig. 110:
Caracasbaai with the Seroe Mansinga at the left side of the picture. Note the almost vertical escarpment marking the rupture of the rock mass, which slided down during the Caracasbaai event. In front, a sequence of shingle beaches has been built up by storm activities.

5.2.1.1 The Probable Genesis of Caracasbaai and the Caracasbaai Slide

Caracasbaai, a bay with an open front of about 1 km in width and a maximum depth of about 250 m, is situated along the south-western side of the island to the west of Spaanse Water (Fig. 110).

Due to its protection against the persistent easterly trade winds, it has been a safe anchorage since the occupation by the Dutch in the 17th century; from that time period Fort Beekenburg remains. Today, a smaller oil terminal is located in the bay. DE BUISONJÉ & ZONNEFELD (1976) first pointed out, that the bay is a rather unusual feature in the coastal landscape and suggested that Caracasbaai owes its existence to a submarine landslide taken place during the Holocene. A huge coastal fragment of about 1 km² (its top surface at least 45 m asl and its basis along the seaward side about 250 m below sealevel) slided down the submarine slope in a rather sudden event in a south-southwestern direction. The volume of the coastal limestone block is estimated at 0.15 km³, representing a weight of some 375 million tons, and the volume of unconsolidated *Globigerina*-oozes, set in motion by the sliding block, is amounted to 0.7 km³ (DE BUISONJÉ & ZONNEFELD, 1976). The submarine scar is recognizable over a distance of several kilometers in front of Caracasbaai down to a depth of about 1300 m (Fig. 111).

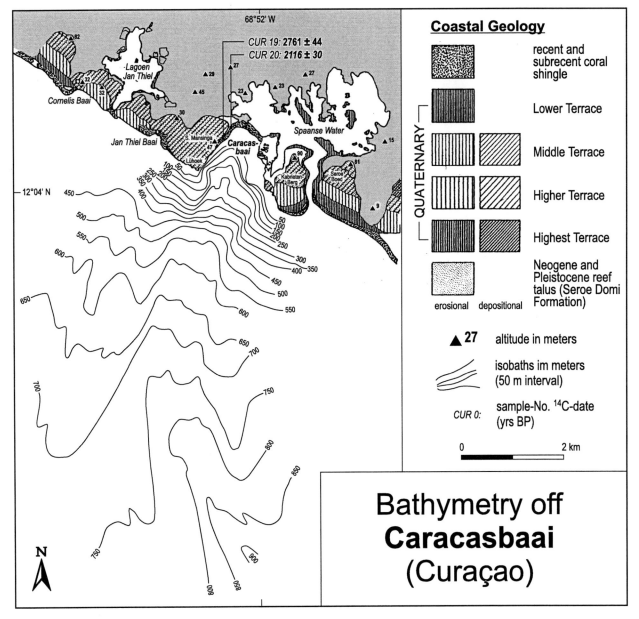

Fig. 111: Bathymetric map of Caracasbaai. The isobaths trace the submarine scar, which is excavated from the sea bottom. The scar is situated about 150 m below the surrounding level of the sea bottom and has steep sides (Source: DE BUISONJÉ & ZONNEFELD, 1976).

DE BUISONJÉ & ZONNEFELD (1976) found evidence by echosounding profiles that the block slides down to 800 m depth as single unit until it broke into smaller fragments. This was confirmed by a submersible dive of the Harbor Branch Oceanographic Institution (REED, 2001), on which a boulder zone, apparently slumped from Caracasbaai, was located in a depth of 700 m (Fig. 112).

DE BUISONJÉ & ZONNEFELD (1976) assume furthermore, that "during the slide a withdrawal of seawater took place, closely afterwards followed by a huge wave, a tsunami, that swept over south-eastern Curaçao and probably inundated a large part of the hinterland" (DE BUISONJÉ & ZONNEFELD, 1976). The authors suggest that the tsunami wave might be responsible for the disappearance of the Lower Terrace along the Caracasbaai itself and from the southwestern side of Seroe Mansinga. Nevertheless, it can not be excluded that the fossil reefs were probably removed and transported down slope during the slide event itself.

In a worldwide comparison of submarine landslides, which caused tsunamis, the Caracasbaai slide can be classified as an event that could have highly likely triggered a local tsunami. A landslide of comparable size close to Nice (France) caused a local tsunami, which killed 11 persons (NISBET & PIPER, 1998). In contrast, giant landslides causing teletsunamis are in the order of some hundred to thousand km^3 like the Storegga slide (Norway), one of the greatest slides with approximately 3880 km^3 and active about 7100 BP (BUGGE, BELDERSON & KENYON, 1988; DAWSON et al., 1988) or the Alika Debris Slides at Mauna Loa (Hawaii) with a volume of approx. 400 km^3, which moved around 240 ka and 105-127 ka (LIPMAN et al., 1988).

Unfortunately, no onshore deposits were found, which could be linked unambiguously to the Caracasbaai slide. The inundation height remains therefore unknown. Nevertheless, it is evident that the slide event is not responsible for the older ridge formations or boulder movements along the leeward coast of Curaçao or the neighboring islands. The age of the older leeward deposits corresponds to the age of the older deposits along the windward coasts of all islands.

It can be excluded that a local tsunami, triggered by the slide event, reached the wave height and the energy to deposit ridge formations of this extent or boulders of the described size and volume along the

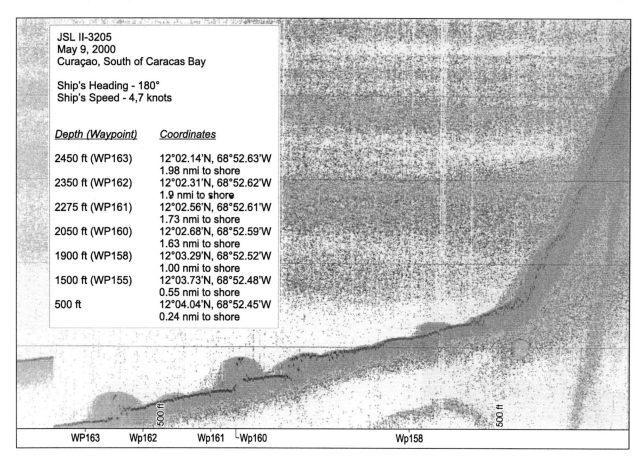

Fig. 112: Fathometer profile, south of Caracasbaai, taken during the submersible dive of the HARBOR BRANCH OCEANOGRAPHIC INSTITUTION on May 9, 2000. At the depth between 700 - 800 m two pinnacles, each up to 60 m high and 550 m in diameter, which slumped form Caracasbaai, have been surveyed (Source: REED, 2001).

opposite (windward) coastlines. The slight dislocation of the huge boulders and boulder clusters (Fig. 113) around Fort Beekenburg is the result of an older movement on top of weathered diabas formations, which acted as a gliding surface on the slope of ancient easterly Seroe Mansinga.

Seroe Mansinga (47 m), at the western side of Caracasbaai, has been cut by the slide with a sharp perpendicular scarp at its eastern side. At the foot of the scarp later shingle beaches with boulders slided down from the steep scarp wall developed. The innermost beach is separated by a small strip of lagoonal water (including some mangrove shrubs) from a central rubble beach deposit at 0.5 m asl, followed by another narrow lagoon and an outer beach with cobbles and rubble. All three beaches have been deposited subsequent to the slide during storm events, as evident by the low height of the deposits and the irregular depositional matrix of the material, which predominantly consists of *Acropora cervicornis* and partly *Acropora palmata* fragments.

Two coral samples for ^{14}C-dating were collected out of the middle beach deposit with a result of 2116 ± 30 and 2761 ± 44 BP, respectively. This data can be considered as minimum ages for the Caracasbaai slide, which of course may be much older. Nevertheless, the coastal geomorphology (e.g. depth of notches in tilted boulders, karstification intensity) in the Caracasbaai area strongly suggests the idea of less than 10,000 years or the Younger Holocene for the event. Coherently the interpretation as given by DE BUISONJÉ & ZONNEFELD (1976) seems a likely assumption, and we exclude a Pleistocene age for the slide event as well and limit the time of the event of less than 4000 years BP.

5.2.2 The Windward Coast (N to S)

Along the windward coast, coarse littoral sediments above low water level are deposited mainly on top of the Lower Reef Terrace. The terrace culminates in the north to a maximum of +12 m asl and lowers down to Oostpunt to +2.5 m. Only about 2% of the windward coastline show gravel and pebble beaches. Sandy beaches are extremely rare and restricted to the wider bokas like Boka Grandi or Boka Patrick, but generally they do not exceed 100 m in width. These beach sediments are mostly carbonates, either from the foreshore environments with corals and calcareous algae, or as abrasive products from the fossil coral reef. Non-carbonates from the basement

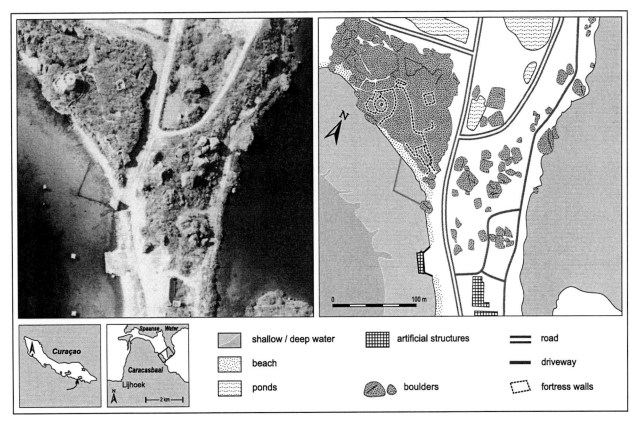

Fig. 113: Overview of the geomorphologic situation of Caracasbaai, south coast of Curaçao. Huge boulders (the smallest with several 1000 metric tons) are located in the innermost part of Caracasbaai. Several clusters of boulders showing their former attachment to a single block mass. Fort Beekenburg, a 17th century fortification is build on one of the blocks.

of the island can be found in lower amounts as a component in the coarse fraction. Beachrocks are widely common in many bokas, but they all are in a stadium of abrasive and bioerosive destruction proving that beachrock cementation is not a contemporary process on the island.

Noordpunt to Hato area

The most striking feature around the northernmost point of Curaçao (Noordpunt) is the sudden change from a perpendicular cliff front to convex cliff profiles with intermittent benches. The reason for the different cliff morphology is the change in exposure: In the sheltered leeward exposure, the bioerosive development of a notch around the tide level is the dominant geomorphological process. Occasionally, it leads to a collapse of the cliff, which provides a fresh steep cliff front. With bending of the coastline into the more windward exposure, the belt of the spray zone, where saltwater spray regularly moisten the exposed limestone surface, becomes wider. The bioerosive activity now reaches high above into the supratidal environment, where extensive biological caused erosion is responsible for the recession of the upper cliff front. Additionally, the rock zone around the fringe of strongest regular surf is protected by accretions of vermetids and calcareous algae against bioerosion. These two processes lead to the windward normal profile. The transition between the leeward and the windward profile around Noordpunt can be observed over a short section of the coastline of less than 50 m extension. (Fig. 114).

The first tsunami deposits of the exposed coast appear as an assemblage of four boulders (78 t, 40 t, 28 t, 12 t) in a distance of 60 m from the cliff front at a height of more than +8.0 m asl (Fig. 115). The two largest ones are overturned as evident by well preserved rock pools, now situated at the basis of the boulders (Fig. 116). Presumably, they are broken out from the upper part of the cliff directly to the north, but bioerosion has transformed the breaking edges significantly. This hints to a relative age of at least several 100 years for the boulder transport process.

At the north-westernmost part of the windward coast an exceptional rubble ridge is located in front of a wide rampart. The ridge (max. height +2.0 m) has steep inward and seaward facing slopes and is accumulated in a distance of 40-60 m from the cliff front (Fig. 117).

The width of the ridge deposit varies between 5-15 m. The material mostly consists of boulders from several kg up to several tons (~3 t). The debris often is of light color as a result of reworking and re-deposition of the coarse sediments by storm surf action. Nature and time of the reworking event remains unclear, but obviously the event was energetic enough to shift pre-existing deposits for several meters to leeward exposing a light-colored belt on the rocky surface of the Lower Terrace in front of the ridge. During the re-deposition, the surface of the limestone terrace had been flattened. Material picked up on the seaward slope is rolled up the ridge and dropped down the advancing slope, where these fresh looking boulders are now covering the deeply weathered dark debris of the older rampart formation. The larger boulders (>1 t) apparently have not been moved during this reworking process, as shown by their dark and weathered surfaces. Characteristic for the ridge is the large amount of well rounded small boulders, whereas in the rampart well preserved *Diploria* heads are remarkable abundant.

Fig. 114: Oblique aerial view of the Watamula-Noordpunt area with the transition between the more sheltered lateral (right) and the windward cliff profile (left). Some huge boulders, a rampart and a ridge-like feature can be seen on top of the Lower Reef Terrace.

Fig. 115:
Huge tsunami boulders at Watamula, Curaçao.

Fig. 116:
The largest Watamula boulder has been overturned and the rock pools of the supratidal are now situated at its base.

Fig. 117:
Reworked ridge as the front part of an older rampart at Noordpunt, Curaçao. In front of the ridge, a band of shelly sand is sedimented, obviously washed out from the basis of the ridge.

The contrast of the whitish color of fresh deposited shelly sands to the brownish color of older sand deposits due to soil formation processes are visible in an area 60 m east of Watamula blowhole. Most likely these white sands were deposited up to a height of about +7.0 m asl during hurricane *Lenny* in 1999.

With the bending of the coastline, the ridge continues in arcuate undulations some hundred meters further, but lowering in heights until it entirely disappears, as well as the fresh sand deposits and the fresh coral fragments. The rampart, however, stretches out as a characteristic feature of Curaçao´s windward coast. In the area of Dos Bokas, a few very well preserved and undisturbed rampart formations of Curaçao have been preserved, only a small belt of mining activities along the seaward flanks disturb the accumulations. They extend for kilometers with a minimum width of 50 m and the inland border, where the material becomes more and more scattered, lays in a distance of around 120 m from the cliff front. On average, the size of the fragments is about 0.5 m in size, but fragments of 1.0 m may occur as well (Fig. 118). The surface topography is undulating with relative height differences of about 1.0 m, which represents also the maximum thickness of deposits in this area.

Figures 119 and 120 illustrate larger ripple marks – a rare evidence, which have been conserved in the rampart formation, in between the surface of the Lower Reef Terrace surface is partly exposed. These ripple marks are the result of a transport movement by extreme strong currents approaching the coastline out of easterly directions. The same effect is documented by the asymmetric deposition of rampart material at both sides of the bokas along the coastline (Fig. 121). In general, debris deposits are absent at the western flanks of the bokas. Here, the water movement was to strong for any depositional processes even of very coarse debris.

Considering the height of the Lower Terrace of more than +8.0 m asl, the entire water column with extreme current velocities must have reached at least this height. A storm-induced wave with a minimum height of ~16 m would have been necessary to reach this level, but as shown above, a wave of this height was never directly observed at any coastline of the world. As a result of the shallow water depth of only 5.0 m in front of the cliff and 15.0 m in a distance of 100-150 m, each wave higher than 12.0 m would break more than 100 m distant from the shoreline. In general, no waves higher than 4.0 m will reach the coastline without formerly breaking (see Fig. 32).

Characteristic evidence along the entire coastline is a sediment-free belt in front of the rampart formations (Fig. 122). Although the surface is very rough with many rock pools up to 1.0 m deep – representing an excellent sediment trap – no coarse sediments have been deposited and trapped in this zone (Fig. 123).

Theoretically, it may be assumed that coarser fragments have been completely weathered by subaerial karst solution or eroded by bioerosion, but considering the required time period of these processes, which lay in the order of several hundred years, this assumption can be excluded. Instead we suggest, that tsunamis have deposited their material further inland, and hurricane wave action were not capable (at least during the last several 100 years) to transport even small debris up to this height.

Along the flanks of the bokas, the cliff morphology is gradually transformed from the seaward convex profile with a broad bench into a thin organic construc-

Fig. 118:
Perfectly preserved rampart near Dos Boka with very coarse debris approximately 70 m distant from the cliff at +8.0 m asl.

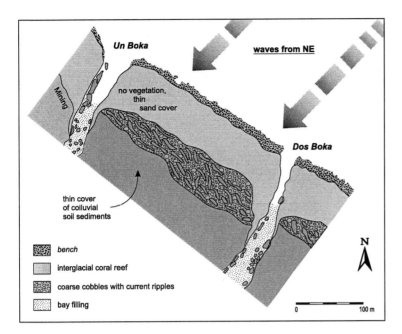

Fig. 119:
Illustration of ripple marks formed in coarse rampart deposits by very strong currents at more than 8.0 m asl, west of Dos Boka.

Fig. 120:
Aerial view of the rampart formation at Dos Boka. Note the ripple marks.

tive platform built by vermetids and calcareous algae (Fig. 124). Finally, the profile is occurring again as the deeply incised notch of the lateral type at the innermost boka sides. Frequently, the flanks of the bokas collapsed and huge limestone boulders with a weight of several thousand tons are left as remnants along the sides. It remains unclear if tsunamis have triggered this process, but it is highly unlikely that the thin organic ledges could have been preserved during energetic tsunami impact. Therefore it can be assumed, that the organic accretions shown in Figure 124 have been constructed subsequent to the last tsunami event, pointing to a relative age of several hundred years for the event.

The origin of the rampart debris was determined via shape analyses of the fragments larger than 20 cm in the area of Un Boka. (Fig. 125).

In total 44% of all fragments are intensively karstified and the original shape is difficult to determine. The remaining debris gives clearer evidence of the source: Less than 1% show the perfect sculpturing of the coral skeleton, from 4.5% the species could be determined. 11% are well rounded. Combining these classes, it can be stated that overall less than 16.5% derive from the foreshore environment. A sharp bioerosive sculpturing form the supratidal rock pool zone could be determined from 40% of the fragments. The remaining 44% karstified fragments mostly show very irregular shapes, which point to the supratidal as the source, as well. This environment clearly has delivered most of the debris in the rampart formation.

It should be emphasized that all fragments derived from the supratidal are of Pleistocene age, as their

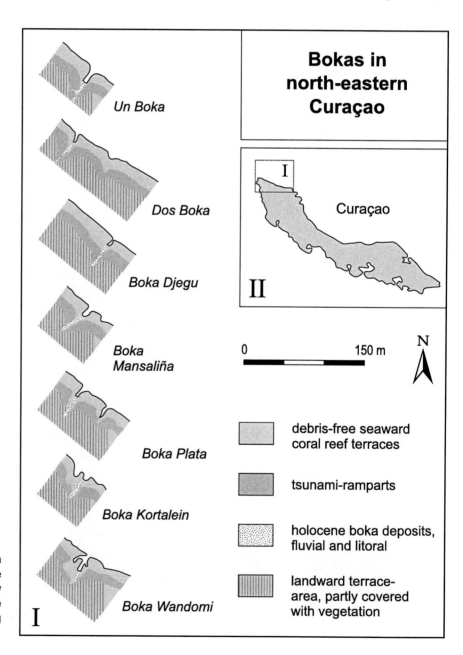

Fig. 121:
The rampart deposition on both sides of nearly all bokas in the northern part of Curaçao show an asymmetric pattern with the nearly absence of debris along the western flanks.

Fig. 122:
Bare karst surface of the Lower Terrace in front of the rampart near Un Boka.

Fig. 123:
Typical rock pools up to 1.0 m deep lacking coarse sediments in the supratidal belt on top of the cliff on the Lower Reef Terrace.

Fig. 124:
A thin organic platform constructed by vermetids and calcareous algae is characteristic for the entrance areas of bokas.

material consists of Lower Terrace limestone, dated back to isotope stage 5 and 7. In contrast, the classes 1 to 3 derived out of the subtidal environment are mostly of Holocene age. That means, over 90-95% of the fragments in this rampart are of Pleistocene age (isotope stage 5 and 7), which will be confirmed in the following measurements of rampart formations. The result evokes consequences for the dating of the ramparts in order that therefore most of the material is not suitable for age determination with the radiocarbon method and the material contain no information about the time of the tsunami dislocation.

The low amount of well-preserved coral species reflect the fact that flourishing reef growth is very restricted along the windward coast. Denser coral growth mainly occurs at the drop off at a depth of -15 m in a distance of 100 to 150 m from the shoreline. Tsunamis evidently have not picked up much material from these areas.

In the undisturbed debris formations west of Dos Boka, situmetric measurements have been conducted at six different sites. The aim was to analyze the orientation of the longest axis of the particles. As shown in Figure 126 at five of the six sites the axis predominantly are perpendicular to the coastline (44-60% in a sector of 60 degrees), representing the direction of the onshore movement. This orientation is especially remarkable at places with good development of imbrication, but due to the rough topography of the ramparts imbrication is not very well preserved. With 45% of the fragments, only one measured site shows a coast parallel orientation.

East of Dos Boka until the area of Hato the ramparts continue, but for the greater part they are completely destroyed by mining. With a more easterly exposure increasing amounts of debris have been accumulated further inland.

This is particular to be seen between Boka Braun and Boka Grandi, where the shrubby vegetation has inhabited the inland parts of the rampart formations. The higher energy impact of this exposition can also be confirmed by single huge boulders, e.g. at Boka Grandi. In an easterly exposition as occurring at Bartolbaai, Playa Grandi and Bergantinbaai, the ramparts are better developed and the size of the single boulders increases. Here, the ramparts are very thick (> than 1.0 m) and they contain relatively coarse debris. South of Boka Patrick and west of Boka San Pedro again huge boulders and extended ramparts are deposited, but unfortunately, they are again extremely modified by mining. Several of the boulders with a weight of >12 t show very well preserved rock pools and bioerosive sculpturing from the supratidal, although they were transported 80-100 m inland (Fig. 127).

In the area of San Pedro the seaward part of the Lower Terrace shows several scarps with a light curvature inland parallel to the cliff front. These pseudo-cliffs are signs of a slow seaward movement of the fossil reef terrace and may indicate the beginning of a slump or even a slide. The instability may be caused by the presence of a deeply weathered base of diabase close to the steep submarine slopes. Along this coastal stretch, the former reef crest of the last

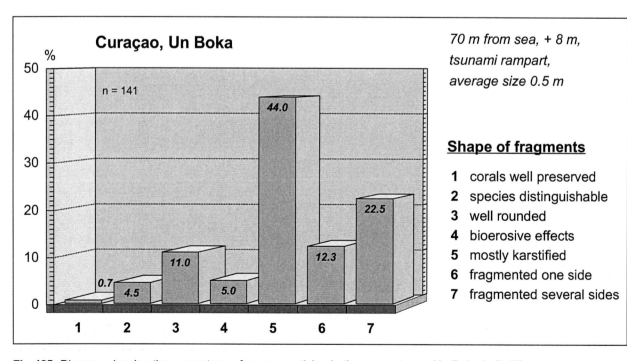

Fig. 125: Diagram showing the percentage of coarse particles in the rampart near Un Boka in 7 different shape classes.

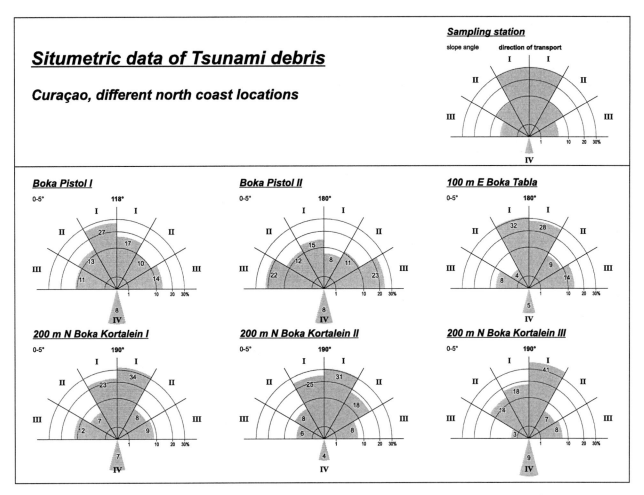

Fig. 126: Orientation of the longest axis of debris fragments on 6 sites in the rampart near Dos Boka, Curaçao.

Fig. 127:
Even large boulders may still show perfectly preserved rock pools and bioerosive sculpturing despite transportation hundred meters inland.

interglacial high sealevel is clearly recognizable. The accompanying lagoon area is still developed as a depression, although at present soil sediments are accumulated here. Inland of the lagoon, the perpendicular cliffs of the Middle Terrace are dominating the landscape. The sharp incised horizontal notch of several meters depth can be followed over several kilometers (see Fig. 14), sometimes with beautiful speleothems, referring to a very limited transformation of coastal forms built more than 100,000 years ago.

Playa Kanoa to Oostpunt

From Playa Kanoa the coastline bends again into an easterly exposure resulting in increased rampart formations and deposition of bigger boulders. Unfortunately, the coastal stretch from Playa Kanoa until some kilometers east of Sint Jorisbaai has been completely destroyed by mining. Nevertheless, this mining is the expression of the former presence of extensive and thick rubble deposits.

The mining shows different techniques: the pattern of the older artificial mining is recognizable in single and chaotic tracks, whereas the modern exploitation uses heavy equipment leading to a complete removal of the debris. In direction Oostpunt several 100 m of undisturbed ramparts are still preserved. The development steadily increases in width, size and inland extension until Awa di Oostpunt. Old and dense shrubs may cover the inland part of the rampart. With increasing size of the fragments the depositional pattern becomes more scattered (Figs. 128, 129). At the easternmost point of Curaçao the rampart formation is reaching more than 200 m inland. As on the northern tip of the island impressive boulder assemblages occur (Fig. 130).

Here, the largest boulder observed on all three islands is located with a size of 5.9 × 5.3 × 3.6 m constituting 281 t in weight. One reason for the excellent development of tsunami depositional features is due to the low elevation of the Lower Terrace with a height of only +2.5 to +3.0 m asl at this location. Additionally, the bench along the exposed coastline of Oostpunt area is interrupted and missing at many places, which hints to a strong destructive impact. Table 13 overviews a list of 86 measured boulders from different sites of Curaçao.

Fig. 128:
Very coarse rampart debris in a scattered layer close to Oostpunt, Curaçao.

Fig. 129:
At Oostpunt the largest tsunami boulder of all three islands can be seen. The weight of this boulder amounts to 281 tons.

At this place, an interesting question concerning the spatial and temporal relation between the coarse rampart debris and the huge boulders should be discussed. Assuming an older age of the ramparts, it should be expected that occasionally boulders have been deposited on top of the rampart debris.

Conversely, an older age of the boulders should be reflected in the depositional pattern of the ramparts due to their influence on currents. However, no evidence for both cases could be observed. It therefore seems to be a reliable assumption that the deposition of the ramparts and the boulders took place simultaneously.

Summarizing it should be pinpointed that in contrast to the leeward side broad and well developed tsunami ridges are absent on the windward side of Curaçao. In the first place, the lack of loose rubble material in the foreshore zone can be regarded as a determining factor, second the lack of a flourishing

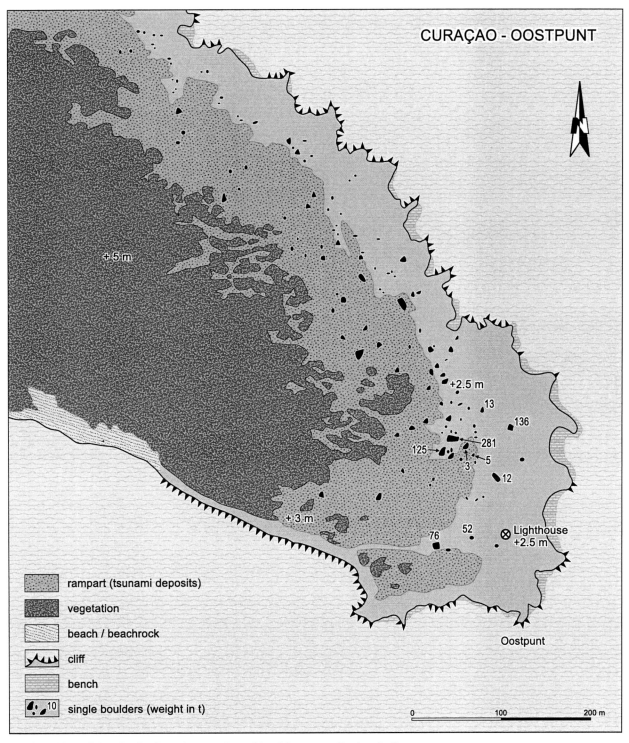

Fig. 130: Map of the distribution of ramparts and boulders near Oostpunt, Curaçao.

Tab. 13: List of large tsunami boulders from different sites along the shoreline of Curaçao.

Location		a [cm]	b [cm]	c [cm]	Volume [m³]	Weight [t]	Height [m asl]	Distance to coastline [m]
Watamula	1	530	100	100	5.3	13.3	5	23
	2	470	340	170	27.2	67.9	5	27
	3	400	200	200	16.0	40.0	5	25
	4	290	120	100	3.5	8.7	5	18
	5	290	270	100	7.8	19.6	5	12
Watamula - Un Boka	6	300	210	70	4.4	11.0	5	10
	7	710	350	100	24.9	62.1	4	4
	8	310	160	120	6.0	14.9	7	36
Un Boka	9	100	100	100	1.0	2.5	6	10
	10	330	190	130	8.2	20.4	10	63
	11	100	100	100	1.0	2.5	6	15
	12	100	100	100	1.0	2.5	6	25
Boka Wandomi	13	310	150	130	6.0	15.1	7	30
	14	200	180	50	1.8	4.5	7	30
	15	210	120	90	2.3	5.7	7	30
	16	240	130	40	1.2	3.1	7	30
	17	260	60	30	0.5	1.2	7	30
	18	200	50	40	0.4	1.0	7	30
	19	260	110	70	2.0	5.0	7	30
Boka Pistol	20	350	220	100	7.7	19.3	4	18
	21	200	130	60	1.6	3.9	4	12
	22	140	110	90	1.4	3.5	4	12
	23	370	320	90	10.7	26.6	4	10
Boka Pistol - Boka Kalki	24	480	350	60	10.1	25.2	6	30
	25	350	220	70	5.4	13.5	6	30
Boka Braun W	26	200	190	140	5.3	13.3	7	70
	27	130	130	80	1.4	3.4	7	30
	28	210	150	70	2.2	5.5	7	25
	29	190	130	100	2.5	6.2	7	25
	30	210	130	90	2.5	6.1	7	30
	31	160	110	90	1.6	4.0	7	30
	32	180	130	90	2.1	5.3	7	30
	33	210	200	50	2.1	5.3	7	30
	34	150	130	40	0.8	2.0	7	30
Boka Grandi E	35	530	290	150	23.1	57.6	5	40
	36	240	110	90	2.4	5.9	5	45
	37	340	340	120	13.9	34.7	5	50
	38	280	190	60	3.2	8.0	5	50
	39	330	140	80	3.7	9.2	5	20
Boka Grandi W	40	200	190	80	3.0	7.6	5	38
	41	260	160	90	3.7	9.4	5	85
	42	200	170	60	2.0	5.1	5	87
	43	250	110	40	1.1	2.8	10	30
	44	130	110	60	0.9	2.1	10	30
	45	130	100	40	0.5	1.3	10	30
	46	190	150	70	2.0	5.0	10	30
	47	150	130	30	0.6	1.5	10	30
	48	340	170	80	4.6	11.6	10	50
	49	230	160	40	1.5	3.7	10	70
	50	320	160	70	3.6	9.0	10	70
	51	210	200	120	5.0	12.6	12	70
	52	250	180	70	3.2	7.9	12	100

Tab. 13: continuation

Location		a [cm]	b [cm]	c [cm]	Volume [m³]	Weight [t]	Height [m asl]	Distance to coastline [m]
Boka Patrick E	53	400	180	80	5.8	14.4	6	7
	54	400	170	120	8.2	20.4	6	5
	55	180	150	100	2.7	6.8	6	60
	56	550	350	120	23.1	57.8	5.5	2
	57	300	200	120	7.2	18.0	9	120
Christoffelpark	58	300	250	100	7.5	18.8	10	80
	59	300	150	120	5.4	13.5	11	100
	60	360	260	140	13.1	32.8	10	80
	61	350	200	100	7.0	17.5	11	160
	62	300	150	110	5.0	12.4	10	140
Slangenbaai W	63	230	180	170	7.0	17.6	3	5
	64	250	180	160	7.2	18.0	3	6
Sheraton Hotel	65	280	280	120	9.4	23.5	2	7
	66	500	420	160	33.6	84.0	2	6
	67	450	230	170	17.6	44.0	2	4
	68	650	460	260	77.7	194.4	2	2.5
Buoy1 W	69	370	200	150	11.1	27.8	2.5	18
	70	500	300	180	27.0	67.5	2.5	20
	71	220	120	90	2.4	5.9	3	20
	72	190	110	100	2.1	5.2	2.5	18
Playa Kanoa	73	380	320	70	8.5	21.3	3	1
	74	800	250	150	30.0	75.0	3.5	25
	75	260	150	140	5.5	13.7	3.5	25
	76	340	220	130	9.7	24.3	3.5	25
	77	400	290	180	20.9	52.2	4	60
Oostpunt	78	540	280	200	30.2	75.6	2.5	60
	79	690	200	150	20.7	51.8	2.5	70
	80	590	530	360	112.6	281.4	2.5	100
	81	500	400	250	50.0	125.0	2.5	110
	82	120	100	100	1.2	3.0	2.5	100
	83	150	120	100	1.8	4.5	2.5	100
	84	200	170	150	5.1	12.8	2.5	100
	85	500	120	100	6.0	15.0	2.5	80
	86	700	390	200	54.6	136.5	2.5	60

coral reef as a source for the debris along this exposed coastline has to be taken into account. And last, the water depth in front of the cliff together with the depositional height of several meter on top of the Lower terrace as present along extended stretches of the coastline require a much higher energy impact than at locations with a lower height of the depositional basis, where ridge features dominate. As a result, on the windward coast ramparts, instead of ridges, are representing the tsunami influence for many kilometers.

Additionally, the formation of ramparts is especially typical for cliff coasts with a well-developed supratidal rock pool area as a main source for the coarse rampart debris.

5.2.3 Summary and Conclusion

Paleotsunami debris deposits are present nearly along the entire coastline of Curaçao, but the accumulations reaching their most extensive distribution as likewise in Aruba and Bonaire along the eastern and northeastern exposure of the windward coast. Whereas usually northern exposed coastal stretches, e.g. Hato area, are free of significant tsunami evidences. The preferential deposition along east-exposed shorelines is documented with the continuation of extensive tsunami deposits along east exposed stretches of the leeward coast of Curaçao. Examples are the ridges and barriers at Awa di Oostpunt, Awa Blanku or along the western section of Bullenbaai. Locally refraction of tsunami waves has led to accumulations even in southern to south-southwestern exposed shorelines like e.g. at Jan Thielbaai or north of Piscaderabaai. The largest ridge deposits of Curaçao are accumulated along the low-lying coast sections in the south of the island, where flourishing shallow fringing reef are developed. The deposition of rampart formations predominantly took place in the northeastern part of the island. Here, steep cliff fronts together with a broad supratidal favor the development of typical ramparts. Mostly they are located in a distance of 20-50 m (seaward margin) to the present coastline. Boulder assemblages of small to medium sizes are characteristic for large stretches of the windward coast, but they are especially notable at the northern and southern tip of Curaçao with weights of over 100 t to nearly 300 t. Nevertheless, even at the leeward coast refracted tsunami waves were capable to dislocate impressive tsunami boulders, e.g. a boulder of 60 t north of Piscaderabaai or even 80 t and 194 t SE of Piscaderabaai.

The overall picture along the coasts of Curaçao gives the impression of very strong tsunami impacts from the east, dislocating far more than 1 Mio tons of coarse debris, from which a greater amount has been removed by mining. Compared with Aruba the impact seems to be stronger, and refraction patterns are better developed. At some places the geomorphology of the tsunami accumulations point to more than one event, which could be confirmed by radiocarbon dating. As on Aruba the field observations suggest that the youngest event took place at least centuries ago. We suggest that inundation has been affected large parts of the island, in particular the low-lying areas around Sint Jorisbaai at the windward coast or Spaanse Water, Piscaderabaai, Schottegat or Sint Marie at the leeward coast of Curaçao. At present most of these places are densely populated and developed for tourism and industrial purposes.

5.2.4 Klein-Curaçao

About 30 km eastward of Curaçao the small island of Klein-Curaçao is situated. It extends only 2.5 km from north to south and 700 m in west to east direction, surrounded by a narrow fringing reef and showing a steep submarine slope profile. The island reaches a maximum elevation of only some meters and is formed by a slightly uplifted younger Pleistocene patch reef. For decades Klein-Curaçao was intensively exploited for phosphates, but still some characteristic features of tsunami debris accumulation are preserved: A rampart along the eastern windward coastline is deposited some meter distant from the shoreline and reaching inland for about 150 m. In the middle part of the east coast, a few large boulders (up to ~15 t) are deposited in a distance of several decameters from the shore. On the leeward side of the island a smaller rampart has been deposited caused by the refraction of tsunami waves with a width of 50 m, but no remarkable boulders could be found (Fig. 131).

Fig. 131:
The narrow rampart along a part of the leeward coast of Klein-Curaçao.

5.3 Bonaire

Bonaire is the easternmost island of the ABC-islands, here both – tsunami evidences and those from hurricane *Lenny* – are the most impressive and extensive ones of all islands. The island has a boomerang-shape, with an east-west exposure in the southern part and a northeast to west and south exposure in the northern part. The small uninhabited satellite island of Klein Bonaire is located 750 m off the western coastline of Bonaire in the shelter from the trade winds. The northern part of Bonaire is characterized by a hilly topography, in contrast to the southern part, which is rather flat, mostly well below +3.0 m asl, and therefore extremely endangered by tsunami inundation. The landscape in the southern

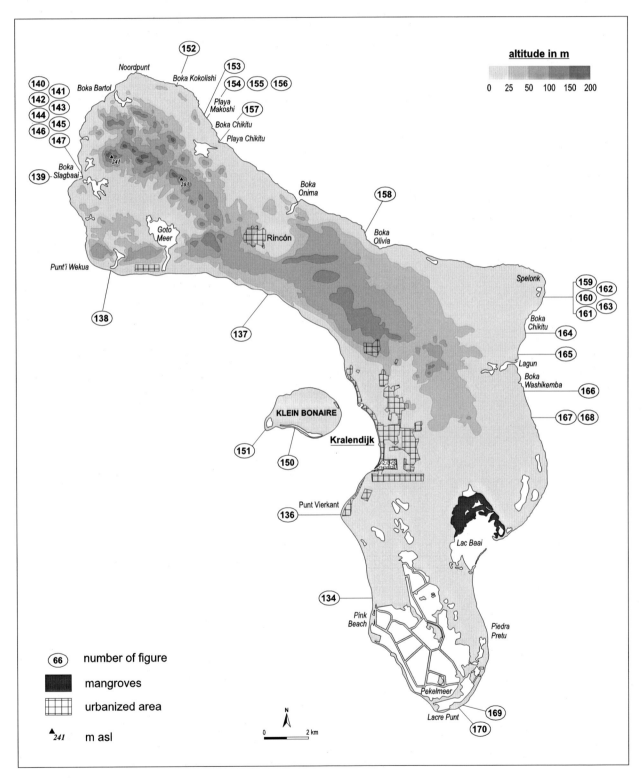

Fig. 132: Overview of the figures described on Bonaire.

part is dominated by salt mining activities in former natural lagoons. Lac Baai at the southern east coast is the largest bay of Bonaire inhabiting extensive mangrove formations. The entrance of this shallow bay is partly closed by small spits and a living coral reef. Like in Curaçao, the coasts of Bonaire are built by the young Pleistocene carbonates of the slightly uplifted Lower Terrace, characterized by steep, often perpendicular cliffs with deep notches. Strikingly, benches as present in Curaçao or Aruba are not very

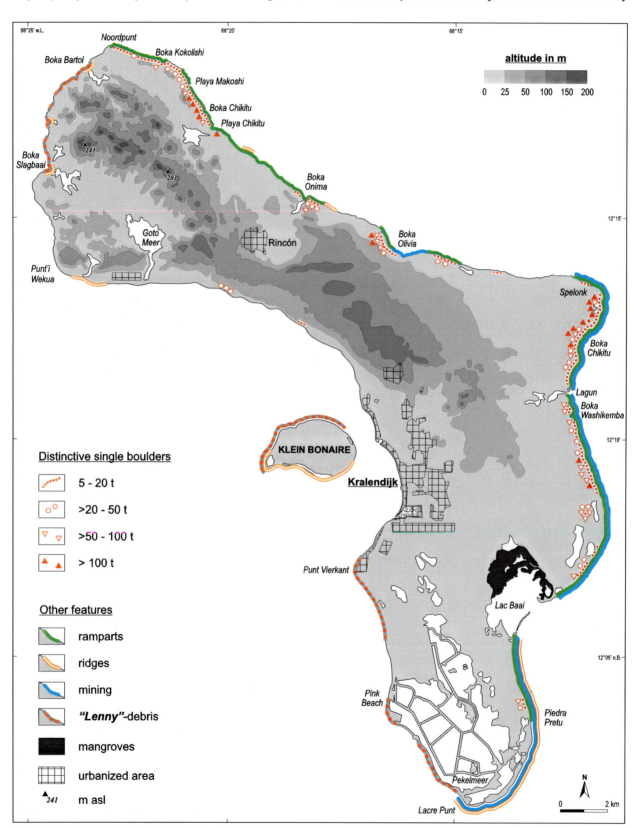

Fig. 133: Distribution of tsunami and hurricane *Lenny* debris along the coasts of Bonaire and Klein Bonaire.

well developed. A few, smaller rubble beach areas can be found (e.g. Pink Beach), but sandy beaches (natural or artificial) are nearly absent. Like on the other islands the infrastructure is exclusively concentrated on the leeward side. Figure 132 overviews the locations of the photographic documents of the tsunami field evidences in Bonaire.

5.3.1 The Leeward Coast (S to N)

The tsunami deposits of the leeward coast are distributed on the coastal section from Lacre Punt to Punt Vierkant in the southern island part, to the entrances of Goto Meer and Salina Tern and in the northern part of the west coast from Slagbaai to Boka Bartol (Fig. 133).

Lacre Punt to Punt Vierkant

The depositional features start west of Lacre Punt on the southern tip of the island with few broken *Acropora palmata* branches and low-lying Lenny-ridges (<1.0 m high) containing mostly of small *Acropora cervicornis* fragments. Deposits are absent around the location Orange Hut, but appearing infrequently again at the height of dive spot Red Beryl as a low ridge. South of Pink Beach (~500 m) an area with extensive mining in younger coral rubble deposits can be seen, but at the broad cape a rather developed and undisturbed *Lenny*-ridge is accumulated. From the location Blue Obelisk to Punt Vierkant a well-developed ridge with sharp inner and outer margins of 1.5 m height extends, the inner edge is arcuate with some tongues of shingle extending leeward (Fig. 134). The deposits are accumulated on top of the Lower Terrace just around the high water mark. At some places, vegetation has been partly buried. Occasionally the ridge splits up in two distinct formations, with the inner one consisting of coarser material. The material is fresh looking, of white color and dominated by *Acropora cervicornis* fragments (90%), but *Acropora palmata* branches or *Diploria* sp. of 0.5 t

Fig. 134:
Fresh looking, wide ridge of coral rubble north of Pink Beach, formed during hurricane *Lenny* in 1999.

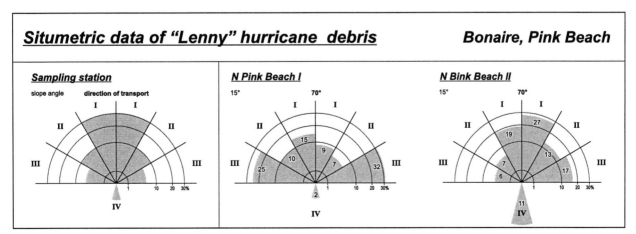

Fig. 135: Orientation of debris on the hurricane *Lenny* ridge at Pink Beach, Bonaire.

are also frequent. A high percentage of the fragments are well rounded due to the movement in the foreshore zone. Situmetric measurements show a rather disordered depositional pattern (Fig. 135). It is evident that the deposition took place during hurricane *Lenny* in 1999 – a single event of low frequency and exceptional high magnitude, which destroyed large parts of the fringing reefs of leeward Bonaire.

Lenny has also left its imprints at the exposed cape at Punt Vierkant, where boulders (1-2 t) broke off the fossil reef edge and were transported 3-15 m inland at a height of only 0.5 m asl (Fig. 136).

Kralendijk to Punt Wekua

North of Kralendijk, the rocky coastline bends to a southerly exposition and the small island of Klein Bonaire additionally shelters the coastal zone. Nevertheless, about 1 km south of Landhuis Columbia the cliff of the Middle Terrace recedes from the coastline leaving a 100 m wide and several 100 m long area of the Lower Terrace open. Here, large boulders in the order of 20 tons are deposited 20-70 m distant from the shoreline at a height of about 3.0-5.0 m. A sheet of *Acropora cervicornis* fragments from the same event are accumulated as well and covering the surface of the Lower Terrace.

It can be excluded, that the boulders have been derived by collapsing of the Middle Terrace, in order their material belongs clearly to the Lower Terrace as evident by the weathering and diagenetic status. Following the coastline to the north the Middle Terrace comes close to the coastline, showing a very well preserved notch from an older (> isotope stage 5) and higher sealevel at around +5 m asl. At Devils Mouth, the cliff front of the Middle Terrace recedes again for more than 300 m from the shoreline. In the center of this fossil embayment, huge tsunami boulders (up to 50 t) appear again on top of the Lower Terrace. East of Landhuis Karpata, an old storm ridge with coarse *Acropora palmata* and *Acropora cervicornis* fragments and a high percentage of rounded fragments is preserved. This kind of old storm deposit is unique for the ABC-islands and was not observed at any other location. Evidently, it is much older than the *Lenny*-ridges. The relative height of the deposit is 1.5 m with two steep slopes on both flanks. From the inland margin arcuate lobes are extending for some meters inland (Fig. 137).

Until Goto Meer the coastal road is constructed on a wider rubble ridge, which might be deposited by a tsunami, but due to the artificial alterations, it is difficult to decide whether the feature is natural or man-made. On Bonaire, even southerly exposed coastlines may show subrecent tsunami ridge deposits as documented by the accumulations at the entrance of Goto Meer and Salina Tern (Fig. 138). The wide barriers of coarse coral rubble (over 0.5 m in diameter) close the small entrances of these rias completely. At the entrance of Salina Tern a ridge of very coarse old weathered debris with a relative height of 2.5-3.0 m and a width of more than 50 m is located. Unfortunately again, the whole area is disturbed by mining activities and no further measurements were carried out. The composition of the material is similar to the described above: *Acropora palmata* and *Acropora cervicornis* branches, but bigger in diameter than the ones deposited in the storm ridges together with *Diploria* sp. and *Porites* sp. The sandy lower basis of the ridge is exposed in an artificial channel cutting the ridge. Estimating the order of the deposited amount it is at least 100 times as much as accumulated in the described *Lenny*-hurricane ridges. The weathered surfaces of the ridge deposits as well as the absence of any smaller rubble deposits from younger storm events clearly confirm that their formation took place in subrecent times.

Boka Slagbaai to Boka Bartol

On the coastal section between Slagbaai and Boka Bartol excellent geomorphological evidences of hurricane *Lenny* are documented in extensive debris deposits. Figure 139 gives an impression of its transportation and transforming capability. Beside that, tsunami deposits in form of ridges are present at the entrance of the smaller bokas.

The extraordinary and multiple effects of hurricane *Lenny* are the result of the unusual track of the circulation system. Hurricane *Lenny* was the first hurricane moving in a west-east direction in the entire hurricane record. Therefore, as the highest waves were sent out of its right front quadrant, their impact stroke the NW-exposed shorelines of the ABC islands directly. The fact that the most impressive sedimentary evidences are found on Bonaire rather than on Curaçao or Aruba can be explained by the successive higher wind velocities the system has gained during its movement to the east (see Fig. 55). As most of the coastline in northwestern Bonaire shows steep cliffs varying from 3.0 to 6.0 m in height, the geomorphologic effects of this hurricane are still rather limited, although the wave height was reported to be in the order of 7 m (pers. comm. Washington Nationalpark Staff, 2001). Consequently, the deposits are more developed in low-lying coastal sections: All existing beaches at the entrances of small embayments or in gaps of the cliff have been filled up with rubble (Fig. 140). At Boka Slagbaai, older tsunami ridges have been covered to about half of their width with the hurricane deposits. Here, *Lenny* destroyed old plantation houses situated at about 2.5 m asl (Figs. 141-143).

Chapter 5.3: Field Evidences – Bonaire

Fig. 136:
Boulders of up to 2 tons slightly moved onshore by hurricane *Lenny* at Punt Vierkant, Bonaire.

Fig. 137:
Old storm ridge deposit east of Landhuis Karpata, Bonaire.

Fig. 138:
The barrier at the entrance of Salina Tern, formed during a subrecent tsunami event.

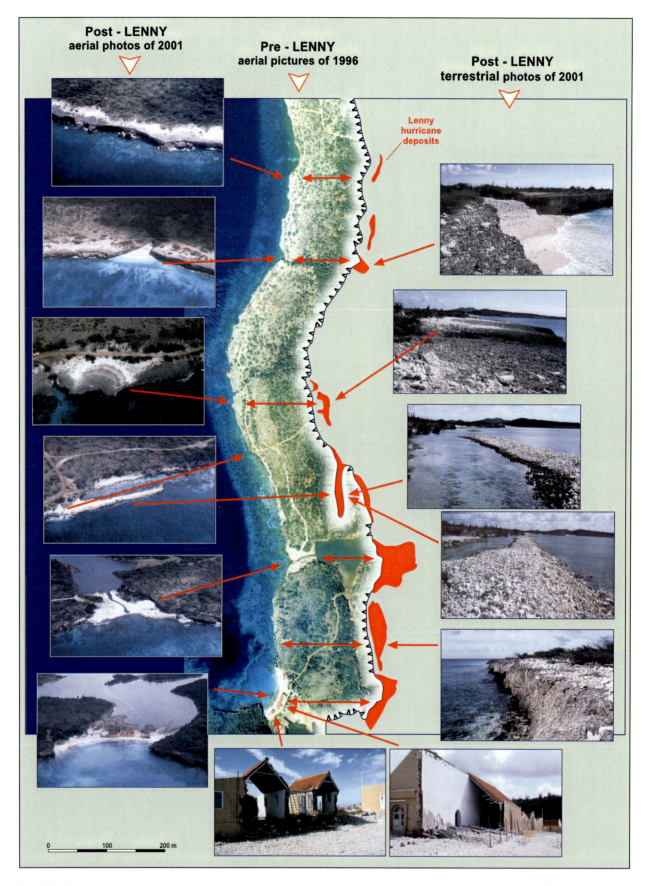

Fig. 139: Evidences of hurricane *Lenny* along a coastal stretch of about 1 km north of Slagbaai. The illustration documents the pre-*Lenny* conditions with reference to an aerial photograph of 1996 and the post-*Lenny* situation by oblique aerial photos and terrestrial pictures (January 2001) as well as mapping results of the onshore deposits.

Fig. 140:
A small embayment in the north of Boka Slagbaai is filled with *Lenny* deposits up to a height of 5.0 m asl.

Fig. 141:
Destroyed houses at the northern end of Boka Slagbaai, Bonaire.

Fig. 142:
High waves hit a plantation house from 1876, destroying the seaward walls until the height of the ridge. Part of the walls were transported further inland.

Fig. 143:
At smaller houses at the southern end of Slagbaai, the direction of the wave impact can be seen: All walls at the seafront were destroyed, and even the backside walls have been collapsed. Only walls perpendicular to the coast have offered resistance and are still in place.

Fig. 144:
Aerial view of the recently formed coral rubble spit by hurricane *Lenny*.

Fig. 145:
Oblique terrestrial photograph of the spit attached to the pre-existing shoreline.

Beside the alteration of existing coastal structures, hurricane *Lenny* created new geomorphologic features as shown in Figures 144 and 145. The most impressive evidence is a new spit of coral rubble with a length of 100 m, about 14 m wide at its beginning and a maximum height from the foreshore basis to the crest of around 4.4 m. The crest itself is highly irregular. An estimation of the volume leads to the assumption that at least 10,000 tons of coral rubble have been accumulated within this feature.

An interesting depositional pattern has been left behind by hurricane *Lenny* on top of a cliff at a height of 5.0 m: The waves accumulated a nearly perfect semicircle of 50 m in diameter of coarse coral rubble (Figs. 146, 147). Coarser fragments (~200 kg) are located closer to the outer margin and smaller pieces more in the inner circle. The particles are very well rounded, which indicate a long stay in the foreshore reef environment, where they have been moved by hundreds of storm waves. The light colored pattern visible in the inner circle can be explained by abrasion and polishing of the weathered surface of the Pleistocene reef terrace by sand and coarser particles during the depositional process.

The general depositional pattern clearly expresses that the wave impact was controlled by the finer topography of the coastline. The semicircle depositional pattern coincides with a semicircle indention in the cliff front. It can be emphasized that depositional patterns of storm-induced wave due to their lower energy level will be much more adapted to the coastal configuration and the physiography of the reef front than tsunami waves.

Fig. 146:
On top of a 5.0 m high cliff a perfect semicircle of fresh coral debris has been accumulated by hurricane *Lenny*.

Fig. 147:
Oblique aerial view of the semicircle debris deposits controlled by wave refraction during hurricane *Lenny*.

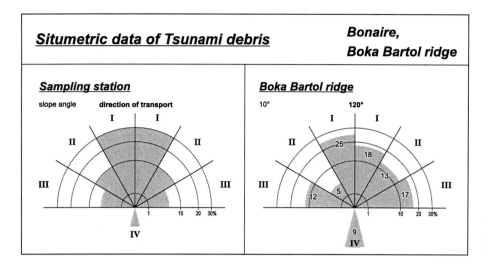

Fig. 148: Orientation of coarse debris on the tsunami ridge at Boka Bartol.

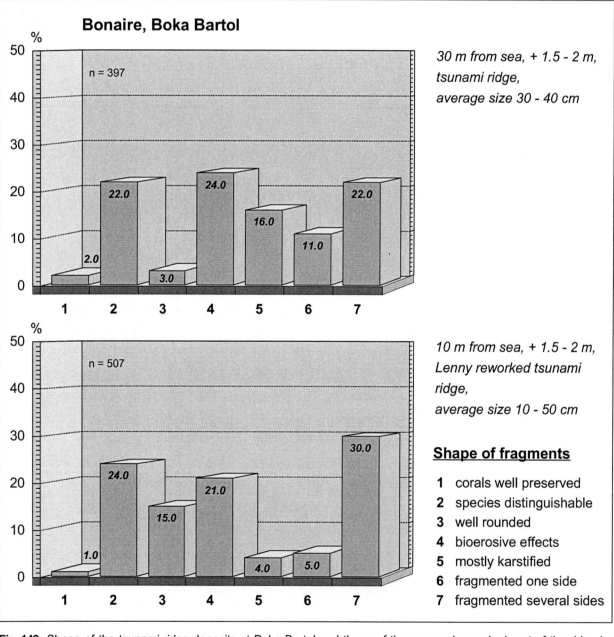

Fig. 149: Shape of the tsunami ridge deposits at Boka Bartol and those of the seaward reworked part of the ridge by hurricane *Lenny* in 1999.

At Boka Bartol, the impressive *Lenny*-deposits are replaced again by well-developed tsunami deposits. Here, only the seaward flank of these subrecent debris accumulations have been reworked and partly covered with recent *Lenny*-material. Situmetric measurements reflect the reworking process partly, the imbrication is still noticeable, but less significant (Fig. 148).

The analysis of the source of the debris with shape measurements at 900 fragments confirms the results of the derivation of material in subrecent tsunami ridges. The measurements were conducted in the seaward reworked flank as well as in the inward undisturbed part. Overall, more than 90% of the fragments in the ridge at Boka Bartol are *Acropora palmata* branches.

As can be seen in Figure 149 we found a remarkable difference in the hurricane influenced part of the tsunami ridge compared to the undisturbed part. The rounding of the fragments is much more characteristic for the seaward margin, whereas overall the rounding is significantly less developed than in other storm deposits. We suggest that at Boka Bartol hurricane *Lenny* mostly reworked fragments, which were already deposited onshore and dislocated rounded debris out of the foreshore environment to a lesser extent.

Another difference is expressed in the amount of weathered pieces with their characteristic pittings on the surface due to karst solution. They are more represented in the inward part of the tsunami ridge, where the wave action from hurricane *Lenny* did not reach the deposits. The border of the wave impact is also documented in the color of the fragments with more whitish fragments seawards and dark, weathered ones in the undisturbed part.

5.3.2 Klein-Bonaire

Klein Bonaire has a west to east extension of 4 km and reaches from north to south about 2 km. The island core consists of the Lower and Middle Terrace deposits. The Pleistocene irregular shape is now completely encircled by a Holocene narrow fringing reef (Fig. 150). Former small embayments were closed by barriers in the south and west, which are built up by subrecent tsunami ridge debris accumulated on top of the reef crest. In width and height they are similar to the barriers of southern Curaçao. The western part of the island has been enlarged by a broad spit, which is nowadays enclosing a shallow lagoon. The width of the barrier consist predominantly of older tsunami rubble, but with bending of the coastline to the north fresh hurricane deposits of *Lenny* are attached in a rather wide belt (Fig. 151). The huge amount of onshore material gives a good impression of the destructive forces of this hurricane on the foreshore reef flats.

5.3.3 The Windward Coast (N to S)

The most impressive depositional evidences of subrecent tsunamis can be demonstrated on the windward coast of Bonaire – not only concerning the great abundance of huge boulders, many of them with a weight of over 100 t – but also with regard to the extension of tsunami ridges and ramparts (see Fig. 133). Except of the southernmost part of the island all coarse debris is accumulated several meters above sealevel on top of cliffs, sometimes with an inland extension of 400 m from the coastline. Table 14 lists the location of measured tsunami boulders on Bonaire.

Noordpunt to Playa Chikitu

From Noordpunt to Boka Kokolishi scattered medium-sized debris fragments are deposited accompa-

Fig. 150:
The southern shores of Klein Bonaire with an old tsunami ridge.

nied for more than 2 km by single boulders of 5 to 20 tons. At Boka Kokolishi beautiful examples of garland-like bioconstructions formed by algal and vermetid rims can be observed (Fig. 152).

They show perfect semicircular growth forms with a height of min. 2.0 m bending against the surf. Inside the boka a small sandy beach is accumulated. The inner flanks of the boka show the same destructive structures of boulders collapsing from the cliff front, as described for the bokas on Curaçao. At Boka Kokolishi, a huge boulder of nearly 100 t (5.5 × 3.5 × 2.0 m) has been disrupted from the cliff front and is now located some decameters distant on top of the cliff (see Fig. 152). Considering the entire geomorphologic ensemble of the boka, it is highly unlikely that a tsunami with the energy to break off and transport boulders of that size has not affected the fragile constructions of vermetids and calcareous algae. Therefore, we assume that these rims have been constructed subsequent to the movement of the boulder corresponding with the last tsunami impact at this place. This event must have taken place some centuries ago, what can be deduced from the time period, which is necessary to construct the algal rims. However, this conclusion should be judged with caution, it still remains open whether this boulder might have been transported over the rims or was derived from a location further seaward from the southern boka flank.

Following the coastline to the south, the exposure changes to a more easterly direction and corresponding to that, the rampart formations are getting more developed and the boulder size increases signifi-

Fig. 151:
The western spit of Klein Bonaire: from the north, the recent storm ridge of hurricane *Lenny* has been attached outside to an older tsunami ridge deposit.

Fig. 152:
Boka Kokolishi with perfectly semicircle grown vermetid and algal rims. Note the boulder of nearly 100 tons on top of the Lower Reef Terrace.

Tab. 14: List of large tsunami boulders from different sites along the shoreline of Bonaire.

Location		a [cm]	b [cm]	c [cm]	volume [m³]	weight [t]	height [m asl]	distance to coastline [m]
Piedra Pretu	1	400	320	180	23.0	57.6	1	15
	2	210	190	130	5.2	13.0	1	30
Boka Chikitu	3	330	330	120	13.1	32.7	4	70
	4	420	250	150	15.8	39.4	4.5	60
	5	500	270	190	25.7	64.1	4.5	100
	6	550	260	120	17.2	42.9	4.5	110
	7	420	200	200	16.8	42.0	4.5	70
	8	400	220	130	11.4	28.6	4.5	90
	9	360	240	100	8.6	21.6	4.5	90
	10	310	200	100	6.2	15.5	4.5	100
	11	300	200	270	16.2	40.5	4.5	100
	12	320	260	150	12.5	31.2	4.5	100
Boka Chikitu Washington	13	850	410	160	55.8	139.4	3.5	25
	14	530	490	220	57.1	142.8	6	80
	15	530	300	190	30.2	75.5	6	100
	16	530	330	200	35.0	87.5	6	100
	17	500	480	200	48.0	120.0	6	100
	20	550	450	260	64.4	160.9	7	50
	21	250	220	220	12.1	30.3	7	50
Spelonk Lighthouse	22	600	450	180	48.6	121.5	4	47
	23	550	240	190	25.1	62.7	4	46
	24	490	370	140	25.4	63.5	4	52
	25	570	320	190	34.7	86.6	4	54
	26	250	250	250	15.6	39.1	4	142
	27	550	260	170	24.3	60.8	4	180
	28	560	320	180	32.3	80.6	4	110
	30	290	220	190	12.1	30.3	4	98
	31	550	250	220	30.3	75.6	4	90
	32	560	190	160	17.0	42.6	4	20
	33	400	250	120	12.0	30.0	4	70
	34	600	370	190	42.2	105.5	4	88
	35	380	260	200	19.8	49.4	4	90
	36	420	200	180	15.1	37.8	4	95
	37	550	250	240	33.0	82.5	4	144
	38	600	340	270	55.1	137.7	4	160
	39	900	350	160	50.4	126.0	4	170
	40	560	260	240	34.9	87.4	4	160
	41	520	260	140	18.9	47.3	4	135
	42	480	320	220	33.8	84.5	4	35
	43	500	410	220	45.1	112.8	4	33
	44	950	350	180	59.9	149.6	4	155
	45	400	370	300	44.4	111.0	4	120
	46	420	350	280	41.2	102.9	4	130
	48	800	350	200	56.0	140.0	4	110

cant. In front of Seru Grandi, with its steep fossil cliff and notch cut in the Middle Terrace, this observation can be substantiated (Figs. 153-156). Nevertheless, before the protection of the area as the Washington Nationalpark, intensive mining has removed all smaller debris fragments, but the large boulders remain in their original position in order they have been too big to transport. Their dark weathered surfaces as well as the karst surface of the debris indicate that the transport process by a tsunami occurred at least centuries ago. In general, all boulders and nearly all fragments of the former rampart formation derive from the active cliff front or the supratidal rock pool area.

At Playa Chikitu, the next embayment to the south, a first impression of an enormous boulder with a weight of several 1000 t seems to prove evidence for an extraordinary tsunami impact. But in fact, the boulder can be undoubtedly classified as a remnant of the Middle Terrace by the occurrence of a deep fossil notch, incised during a sealevel highstand of the last interglacial period. Directly south of Playa Chikitu, excellent rounded cobble-sized rubble have been accumulated out of the reach of modern storm surf (Fig. 157). Within the deposition at least two different units may be distinguished, with the older one strongly cemented. Close to this deposit, one of the largest tsunami boulders of Bonaire with a longest axis of 9.20 m and a weight of 120 t was perched.

Playa Chikitu to the North of Spelonk
Along the coastal stretch from Playa Chikitu to Spelonk the windward shoreline mostly has a northerly exposure. Here, a broad rampart continues until Boka Onima, where large boulders – one of them around 50 t – are situated on the southern side. Some 100 m to the southeast, a rather high and steep sloped tsunami ridge is developed – a unique feature along the entire windward coast of northern Bonaire. At Boka Olivia, an assemblage of six huge boulders is deposited in a distance of about 60 m from the cliff front at height of +5.0 m asl The weights of the largest ones amount to 180, 125 and 80 tons, respectively. One (125 t) seems to be broken apart during the deposition process (Fig. 158). The rampart material now is thinning out, before it disappears entirely further to the east.

Spelonk to Lac Baai
The coastline bends again to a directly eastern exposure at Spelonk Lighthouse. From here stretching for about 10 km south the most spectacular rampart formations and boulder assemblages of subrecent tsunami events are documented, although mining activities have extracted approximately 1 Mio. t of small to medium-sized debris, leaving only the largest as remnants behind. On this coastal stretch, the cliff front is only about 3 to 4 m high and the surface of the Lower Terrace reaches +5.0 m asl. A bench is either not developed or has been destroyed. Not only the amount and size of the debris, but also the extension of the inland deposition reaches high numbers. Rampart and boulders are transported as much as 200 m, occasionally even up to 400 m inland – the record with respect to all three islands.

Figure 159 depicts aspects of the most inland part of the boulder field near Spelonk Lighthouse: The boulders are situated within well-developed vegetation of wind-formed shrubs, some are even growing on top of the boulders. Initial soil formation took place in places around the boulders. In congruence with the rough karst topography of the Lower Terrace surface, which partly has sculptured the coral fragments out of the intercalating cements, these observations suggest a rather old depositional age. Figure 160 shows some of the largest boulders, with their longest axis orientated parallel to the coastline.

Two reference areas have been mapped in large scale in order to illustrate the distributional pattern of the tsunami deposits and to document the influence of mining (Figs. 161 and 162).

Twenty-seven boulders with a weight >10 t were mapped in the northern reference area (see Fig. 161), 9 of them having a weight over 100 t. At least two of these have been transported over a distance of 150 m from the cliff front. One boulder of initially ~263 t broke apart during the moment of deposition as obvious by coral fragments in both boulders, which fit exactly together. Now the boulder is separated in two fragments of 125 t and 137 t (Fig. 163). Some boulders show nicely preserved rock pools at their top surface, giving the idea of a transportation process rather swimming than rolling or tumbling (Fig. 164).

The map of the southern reference area (see Fig. 161) shows 147 boulders with weights >3 t, among which 13 were measured. The tsunami impact was reduced in this area as concluded from the smaller size of the boulders and the existence of a partly existing bench.

Some general findings should be pinpointed: The Spelonk area is remarkable in the sense that extreme huge boulders and broad rampart formations are present in the depositional record. Further south (area Lagoen) the boulders decrease in size whereas the ramparts are becoming even more developed in their horizontal and vertical extent.

What is the reason for this spatial differentiation? Boulder assemblages are coinciding remarkably often with coastal sections, where a bench is lacking and the cliff front is nearly perpendicular. Here, a tsunami breaks off boulders out of the upper part of the cliff. This coastal configuration also shows mostly a rather narrow supratidal zone, which lacks a developed and mature rock pool zone, so that only small amounts of debris material can be derived from this environment. If the coastal physiography leads to the development of a rather broad supratidal with a more convex cliff profile, the amount of debris will increase significant as more material can be derived by a destructive impact. Consequently, rampart formations are more likely to develop.

Nevertheless, the identification of the fractured surface areas is in any case rather difficult, because of the subsequent fast transformation by bioerosive processes.

Chapter 5.3: Field Evidences – Bonaire

Fig. 153:
Oblique aerial photograph of large boulder assemblages in front of the Middle Terrace of Seru Grandi in Washington Nationalpark, Bonaire.

Fig. 154:
Boulder of more than 100 t located in Washington National Park in the area of Seru Grandi.

Fig. 155:
Boulder of more than 100 t deposited 100 m distant to the cliff front.

133

Fig. 156:
Assemblage of two boulders with a weight of about 100 t.

Fig. 157:
Partly cemented tsunami rubble on top of a cliff south of Playa Chikitu. In the distance, a huge tsunami boulder 9.20 m long with a weight of 120 t is visible.

Fig. 158:
Boulder assemblage at Boka Olivia, one presumably broken apart during transportation and deposition.

Regarding the rapid progress of bioerosive overforming in particular the supratidal environment by boring and grazing organisms reaching more than 1 mm/yr or even increased in weakly cemented young limestones (KELLETAT, 1985; TRUDGILL, 1987), it can be calculated that within 100 years the erosion will remove about 10 cm (max. 30 cm) of the surface layer. As a consequence in a time period of only some 100 years no signs of breakage or surface fracturing will be preserved.

Further to the south extreme mining has removed at least 90% of the rampart material and the devastating impact for the coastal landscape of the often uncontrolled exploitation can not be overseen (Figs. 165, 166). The straight and curved artificial depositions in the mining areas in Figures 165 and 166 are mainly remaining *Acropora palmata* branches with several 100 kg in weight. They are evidently not suitable for construction purposes due to their size. Likewise, single large boulders often up to ~50 tons

Fig. 159:
An impressive boulder field south of Spelonk Lighthouse, situated in rather dense vegetation more than 150 m apart from the shoreline.

Fig. 160: Four examples of huge boulders deposited in the area south of Spelonk, each with a weight of over 100 t.

PALEOTSUNAMIS IN THE CARIBBEAN

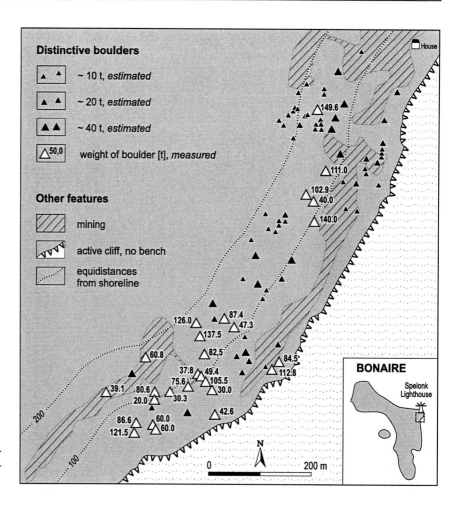

Fig. 161:
Detailed map of the northern reference area 3 km south of Spelonk Lighthouse.

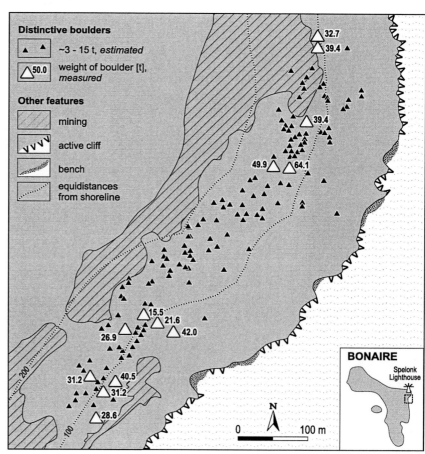

Fig. 162:
Detailed map of the southern reference area at Spelonk Lighthouse.

Fig. 163:
These two boulders of 126 and 137 t have been moved by the tsunami in one piece and settled down about 160 m distant to the cliff at a height of +5.0 m asl. During the deposition, the boulder was broken and the fragments separated for several meters. The breaking of the two pieces can clearly be demonstrated by the fitting of large coral colonies in both of them.

Fig. 164:
Very well preserved rock pools at a boulder 120 m distant to the shoreline.

Fig. 165:
Extensive mining has removed most of the debris and exposed the unweathered surface of the Lower Terrace, which show initial soil formation before the tsunami events documented with the brownish colors.

Fig. 166:
Heavy equipment in modern times leads to complete devastation of the landscape and a more accurate pattern of the residual deposits.

Fig. 167:
Sharp contrast between old undisturbed rampart and fresh mining areas south of Lagoen.

Fig. 168:
The ramparts south of Lagoen clearly show two parallel units with slightly different color and therefore presumably different age.

remain in situ, and are now located scattered in a rather bare landscape. Two boulders in this area have an estimated weight of more than 100 tons.

In Figure 167 the front of the mining process is bordering against an undisturbed rampart south of Lagoen. The undisturbed deposits may represent two different depositional events, as suggested for the two or three parallel debris ridges on the leeward side of Curaçao. South of Lagoen, the ramparts give as well the impression of two separated units: One rampart accumulated more inland with a light gray color and a less dense packing arrangement of smaller sized debris. In contrast, the seaward rampart shows a dark gray color, coarser fragments and an increased vertical height. The packing arrangement is denser and a strip of ripple marks can be observed at the front edge (Fig. 168). The age of the two units may differ about several hundred to thousand of years, as will be discussed in Chapter 6.

Close to an old plantation house near Boka Washikemba the most far-reaching deposition of tsunami boulders can be observed. Three boulders of more than 10 tons have been deposited 350 to 400 m distant to the coastline beyond the mining area. The dimension of the mining area hint to an extent of a former rampart of approximately 300 m. Supposedly the vertical dimensions thin out with increasing inland extension. An old wind-shaped Divi-Divi tree grow in the opening of Boka Washikemba at only 2.0 m asl (see Fig. 6). The age can be estimated of ~300 years or even older, suggesting a last tsunami impact prior to its growth. The described mining and depositional pattern continues in direction of Lac Baai, although in less accentuated dimensions.

Fig. 169:
The broad and coarse tsunami ridge deposits along the southeastern coastline of Bonaire.

Fig. 170:
Seaside vertical section of the tsunami ridge in southern Bonaire showing a weathered coarse upper layer with fresh looking sand at the basis.

Lac Baai to Lacre Punt

South of Lac Baai the coastline is characterized by a more convex cliff profile with a broader supratidal belt. A fringing reef is developed offshore. Because of the lack of a cliff but with a fringing reef developed the process of tsunami destruction and deposition changes significantly. For some kilometers a ridge appears, gradually getting broader and higher to the south where the reef is better developed and the foreshore becomes shallower and wider.

This foreshore environment may supply not only sand and coral rubble reworked by light and medium storm or hurricanes, but also living reef colonies, which do not withstand severe impacts. The impact of tsunami waves result in the deposition of an extensive ridge deposit (Fig. 169).

Over a distance of several kilometers the ridge reaches a thickness of 2.0 to max 3.0 m with a rough undulating surface topography of about 1.0 m. The ridge is presently about 50 m wide, but it is extensively disturbed and removed by mining at its inland margin. By far most of the fragments with an average size of 30-50 cm are well rounded and imbrication is common. The seaward part of the ridge is reworked by recent surf impact and the fresh exposures show a congruent orientation of long axis, deep weathering and fresh looking sand and shells at the basis (Fig. 170). These fine sediments have been washed out from the upper layers. In no case younger storm events had been able to deposit fresh fragments on top of the old tsunami ridges. On six sites situmetric measurements have been conducted (Fig. 171).

Beside one site, all of them showing an orientation of the longest axis predominantly perpendicular to the coastline, which reflect the direction of the tsunami current. This coincides with partly well preserved imbrication of the fragments reflecting the same current direction. Only in areas of a chaotic topography a rather indifferent or coast parallel orientation may be observed.

About one km east of Willemstoren Lighthouse more than 1000 fragments have been determined regarding their shape and origin (Fig. 172). Only very few fragments show a good conservation of the fine-structure of the coral skeleton, but still a high percentage (16%; 21%) can be identified, in particular the typical *Acro-*

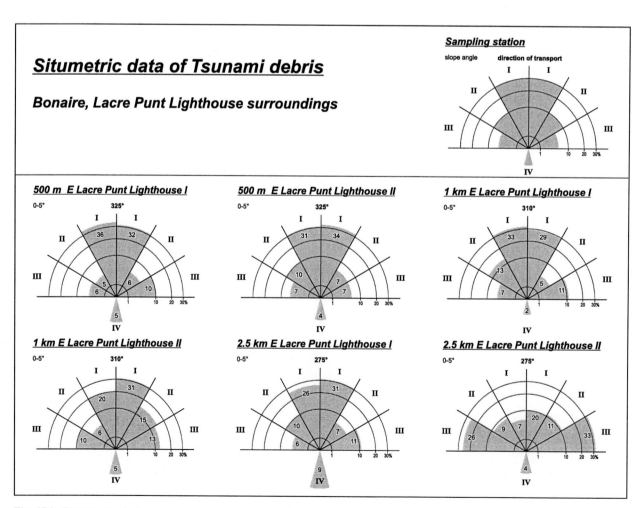

Fig. 171: Diagrams of situmetric measurements, southeastern Bonaire.

pora palmata branches. Well-rounded fragments amount to 13% resp.14%, those with significant bioerosion contribute with 30% and 23%. Broken branches are common, mostly with a fragmentation on more than one side. A karst surface is mainly well developed on 14% resp. 13% of the debris fragments.

Summarizing the results, it can be stated that the debris fragments are originating mainly from a flourishing *Acropora palmata* reef growing in a short distance offshore, but many particles stayed for a longer time period in the foreshore environment as evident by rounding and bioerosive borings. The similarity in size, shape and orientation indicate one depositional tsunami event, as will be proved by ^{14}C-dating (see Chapter 6).

5.3.4 Summary and Conclusion

Bonaire exhibits the most extensive paleotsunami deposits of the described depositional types – ridges, ramparts and boulder assemblages. Both, the position at the eastern end of the Lower Lesser Antilles island chain as well as the predominant east-exposure of its windward coasts, have favored the development of broad tsunami ridge deposits along the southeast coast, in particular around Willemstoren Lighthouse, as well as wide tsunami ramparts all along the east coast. The occurrence of impressive boulder assemblages on the east coast between Spelonk Lighthouse and south of Lagoen, at Boka Olivia, or near Seru Grandi in the northern part of the island underlines these evidences. Here, weights of more than 100 tons, which are deposited

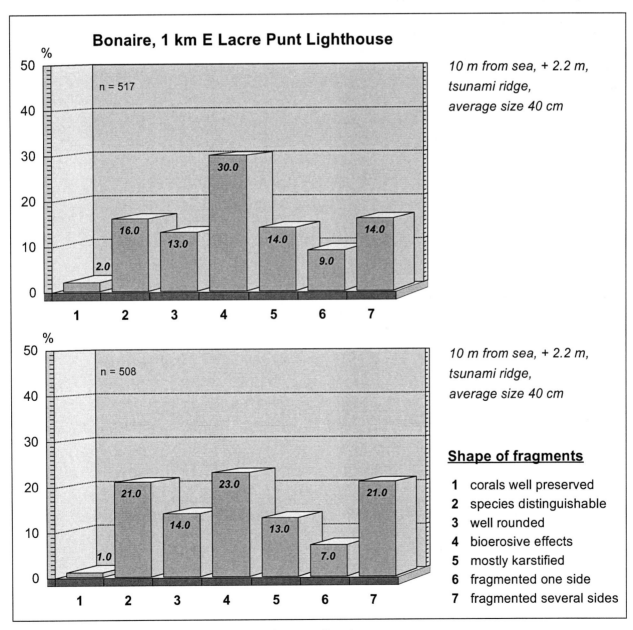

Fig. 172: Diagrams of shape measurements, southeastern Bonaire.

far inland are not exceptional. The distribution of paleotsunami deposits strongly suggests an impact out of an easterly direction. Refraction of easterly tsunami waves with accompanying depositional patters can be observed at ridge deposits at the entrance of Goto Meer as well as in the same exposure at the south coast of Klein Bonaire.

Additionally, the geomorphologic evidence of hurricane *Lenny* (1999) is excellent documented on the leeward coast of Bonaire, in particular in northwest to west exposures, as well as along the northern coast of Klein Bonaire. These recent hurricane deposits exhibit a striking contrast to the subrecent tsunami debris in state of preservation, weathering, as well as considering the spatial relation to the modern surf environment. A comparison of both impacts with regard to the preserved geomorphological and sedimentological imprints highlights that the scale of the paleotsunami event have been in the order of 100 times stronger than the youngest effects of hurricane *Lenny*, a category 4 hurricane system.

It can be assumed, that the southern part of Bonaire, including the area of Kralendijk and the airport area, due to the low relief have been inundated during each of the (at least) two or three tsunamis of the younger Holocene.

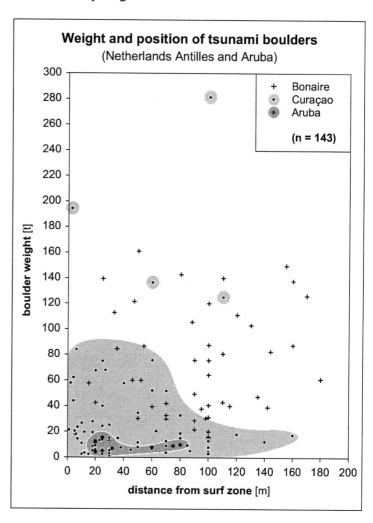

Fig. 173:
Weight and transported distance of boulder deposits for the ABC-islands.

5.4 Comparison of the Paleotsunami Deposits of Aruba, Curaçao and Bonaire

The studied field evidences of paleotsunami deposits cover the region extending from Aruba to Bonaire over a distance of about 200 km. Similar physiogeographical settings regarding climate, geology, coastal geomorphology and land- and coastal ecosystems characterize the islands. Nevertheless, the islands show a distinctive differentiation in the spatial distribution, the extent as well as the amount of coarse paleotsunami deposits. But the depositional character of the debris formations in form of ridges, ramparts and boulders (single or assemblages) remains the same and recurs on each island: The broad tsunami ridges have been accumulated close to the coastline and are more present at places with low lying fossil reef platforms with a convex profile.

Chapter 5.4: Field Evidences – Comparison

The material derives predominantly from the subtidal. The ramparts with coarser debris fragments occur in broader belts, always decameters distant to the modern surf zone. They are deposited at a higher altitude on the coastal platform and mainly concentrated on the windward coastlines. The material originates to a large amount out of the supratidal. Boulders, single or in assemblages, are perched in front of or occasionally within the seaward margin of the ramparts. They are broken out off the cliff front or the bench. All these coarse sediments are rather old and weathered and not to be associated with young storm events. At some places clearly two or three phases of deposition can be distinguished. The field evidences strongly suggest one or more simultaneous impacts on all islands. This observation was proved by dating of the debris fragments in Chapter 6.

Figure 173 summarizes the results of the boulder distribution in weight, transport distance and depositional height for all islands. The differences in boulder weight on the islands expresses a first clear picture: Only nine remarkable boulders could be found on Aruba – all of them rather small with a weight <20 tons. On the largest island Curaçao, in total 86 impressive boulders along the coastline were chosen for further analysis- beside four boulders, all with a weight of less than 90 tons. Here also, the two largest boulders of all islands were deposited; one (194 t) on the leeward coastline and another one (281 t) on the most exposed southern tip of the island. For Bonaire, overall 48 impressive boulders were selected for detailed measurements because of their remarkable size – thirteen of them >100 tons.

The graphic clarifies that in general on Bonaire more boulders in higher weight classes compared to the other islands have been deposited by paleotsunami events. The differences in transport distance reflect the same evidences. The inland extension of boulder distribution reaches the highest numbers with 180 m on Bonaire and decreases from 160 m on Curaçao to only 80 m on Aruba. It should be kept in mind that the illustration represents only the values for the boulders measured in detail, besides unmeasured boulders of significant size may be found further inland. The comparison in the deposition height of the boulders shows a different picture: Now, on Curaçao boulders were deposited up to 12 m asl, whereas on Bonaire and Aruba the height values reach only max. 7.0 m asl This evidence can be explained by the height of the Lower Terrace as the basis for all depositions and which is the highest on Curaçao.

The average weight of the largest boulders increases from Aruba with 9.1 tons (9 boulders) over Curaçao with a mean weight of 88.8 tons for 10 extreme boulders to Bonaire with an average weight of 135.1 tons for the 10 largest boulders (Fig. 174).

Combining the factors boulder weight and depositional distance from the coastline the result can clearly be assigned in groups each representing one island (Fig. 175). Associating all three important factors to energy values, it can be clearly highlighted that Bonaire shows the strongest impact of paleotsunamis and the geomorphological effects are decreasing towards Aruba (Fig. 176).

The origin of the deposits and their classifying to the distinguished debris formations is similar on all three islands. The foreshore environment (subtidal) delivers coral fragments and reworked sediments of different size. This material predominantly built up the tsunami ridges with a percentage of 90% of all fragments, whereas in rampart formations the value decreases to only 10%. In ramparts the fragments derived from the supratidal are dominating. The supratidal together with the intertidal environment with its bench are a possible source area for the boulders of different size classes. But most of the boulders derive from the cliff front, in particular when the cliff is nearly perpendicular.

To dislocate and transport boulders of the discussed size undoubtedly a highly energetic tsunami wave impact accompanied by high current velocities is necessary. The subsea topography with the deep water basins around the ABC-islands favor the conditions for such an impact. Figure 177 illustrates an estimation of a tsunami approaching Curaçao from a N-E sector in compliance with the real water depth. The tsunami will reach Curaçao from 14 km distance (=2000 m isobath) within 2:55 min. The velocity will decrease from 504 km/h to 150 km/h in 700 m distance and further to 44 km/h in 15 m depth. For the last 700 m a travel time of only 35 seconds is needed. In reality this velocity will be higher due to the volume of the water mass running onshore.

Figure 178 represents a synopsis of the most impressive paleotsunami relicts of all three islands.

Bonaire as the most eastern island shows a preference of deposition along the east-exposed coastline. In contrast, the most western island Aruba located on the Venezuelan shelf shows only limited extension of tsunami deposits. Curaçao additionally display ridge and boulder depositions on the southern shorelines, beside extensive evidences on the northeastern coastline. Whether this is the result of a less of sheltered exposure of the leeward coastline compared to Bonaire, or due to a more ESE direction of the approaching tsunamis is debatable.

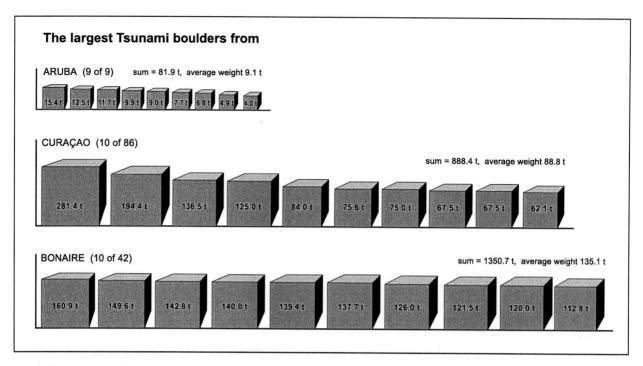

Fig. 174: Average weight of the largest boulders on Aruba, Curaçao and Bonaire.

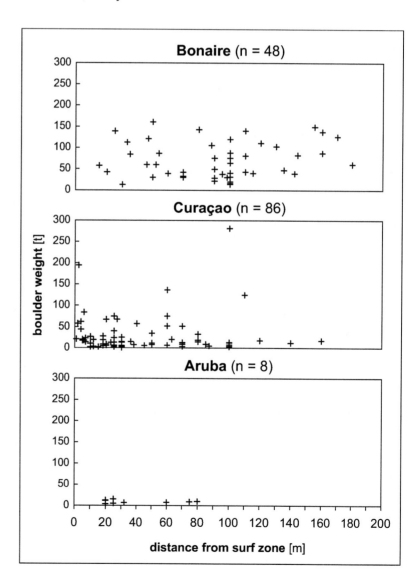

Fig. 175:
Diagram of boulder weight and depositional distance for boulder deposits of the ABC-islands.

Chapter 5.4: Field Evidences – Comparison

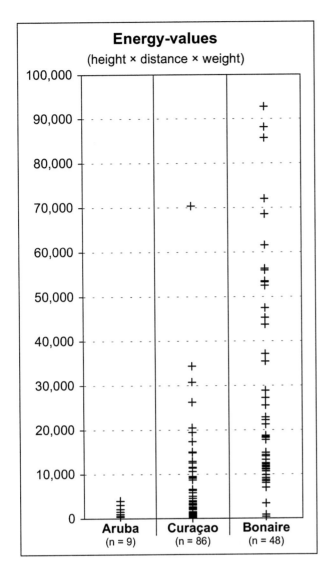

Fig. 176:
Diagram representing the energy level of the impact derived from boulder weight, transport distance and depositional height.

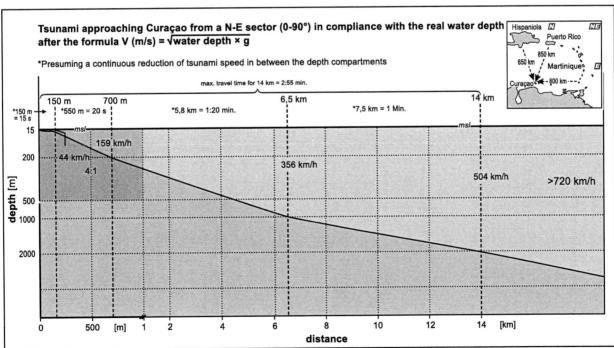

Fig. 177: A tsunami approaching Curaçao. The arrival times reflecting the real values for water depth around the island.

145

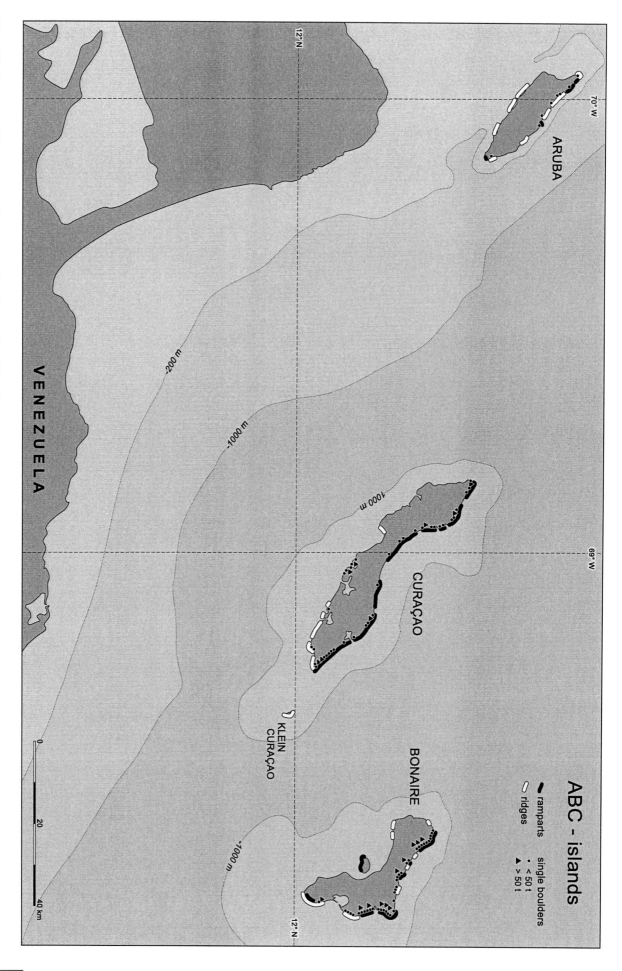

Fig. 178: Overview of impressive paleotsunami imprints on the ABC-islands.

6. Dating and the Depositional History of the Paleotsunami Events

The most important dating instrument in geosciences is stratigraphy. However, this tool is rather limited for deciphering the geological history of coarse debris depositions as the accumulations are usually deposited on the surface without any interpretable context to other strata. Like in the case of the ABC-islands, the depositional basis for the tsunami debris is the Younger Pleistocene coral reef terrace, from which only a maximum age ranging for the depositional event can be conducted. Since no other sediments have been subsequently deposited on top of the (last) tsunami material, stratigraphy alone does not provide further chronologic resolution. Therefore, Radiocarbon dating is applied in combination with several other independent methods derived from field observations to determine the depositional history of the tsunami debris formations.

The key questions in any paleotsunami study are: "How old are the deposits?" and "How many events we see within the geomorphologic evidence?" Before we approach these questions with relative and absolute age dating techniques, some considerations concerning the frequency of tsunami events as can be deduced from the field studies should be discussed. We documented three different types of coarse debris formations for tsunami impacts on Curaçao, Bonaire and Aruba: Ridges, ramparts and boulders. This does not mean that each of these deposits represents a single tsunami event, they might rather be deposited contemporaneous and replace each other in different coastal environments as shown above.

The ridge deposits show occasionally clearly one or more phases of deposition in form of two or three parallel ridges (e.g. west of Oostpunt in southern Curaçao). This may prove two or three different tsunami events, but they may as well represent two or three different waves during the same event. It is evident that a second wave or a subsequent occurring tsunami at the same location will have contact to the existing ridge – reworking, restructuring, removing or just overrunning the old deposit. Consequently, only with field observations it is nearly impossible to differentiate the frequency of tsunamis on the basis of the ridge deposits. Similar considerations are applicable to the rampart formations.

At some places (e.g. S of Lagun, Bonaire) the impression in the field suggests more than one event. But the significant state of weathering and the lack of finer stratigraphic separation give a rather unclear picture. The interpretation of the huge boulder deposits remains equally difficult. Are these boulders deposited before, subsequent or contemporaneously with the smaller-sized debris? Assuming an older age, the depositional pattern of the smaller debris fragments around the boulders should reflect the existence of the boulders in a certain distinguishable way. If they were younger, it would be expected that at least at some places the boulders would have been deposited on top of pre-existent ridges or ramparts. However, none of these scenarios could be observed in the field. The field evidence near Seroe Colorado, Aruba, rather suggests a simultaneous deposition. Here, larger boulders up to several tons in weight are incorporated into a sandy double ridge. In general, the observations in the field indicate that the boulders are deposited together with the smaller fragments, in particular within the rampart formations. Mostly they are deposited in front of the seaward rampart margin, which can be explained by their increased size and weight compared to the smaller rampart fragments.

It can be stated that without absolute age dating methods the frequency and age of coarse paleotsunami impacts may not be estimated correctly, but relative age dating approaches are an indispensable tool to establish a historical tsunami record and to crosscheck the absolute data results. The age of the deposits will be first discussed in the light of relative age indications, before the absolute data will be presented.

6.1 Relative Age Dating

Dating of coarse sediments is a difficult task since no stratigraphical sequence can be interpreted and no analysis methods of sedimentology can be applied in coarse sediments. Nevertheless, relative age indications allow a good estimation of the time range for the minimum and maximum age of the deposits as will be presented with the following conclusions:

- The sediments are stratigraphically the youngest deposits in their environment. No other accumulations took place subsequently;

- In the time span after the deposition, the debris formations were not disturbed by any other natural impact. Only the seaward margin of some ridge deposits close to the shoreline have been reworked by younger events;

- Important relative dating possibilities of a tsunami impact are related to the preservation of bioerosive and bioconstructive coastal features (see also KELLETAT & SCHELLMANN, 2001 a, b). Estimations of the time period needed for the forming processes (bioerosion: ~1-2 mm/yr; bioconstruction: ~2-5 mm/yr) can limit the time range for the event relatively accurate. Transferred to the ABC-islands, it can be stated:

- No signs of fresh outbreaks of limestone material either in the cliff front, the bench or the supratidal zone were found, so that the origin of boulders could be unambiguously identified. Subsequent bioerosive processes made the breakouts unrecognizable, indicating a minimum dislocation and depositional age of at least some hundred years;

- Limited bench development along coastal stretches with major tsunami impact point to several centuries without further impacts of tsunamis, again suggesting an age of some hundred to thousand years;

- The karst weathering of the debris fragments and boulders can be used as another reliable indicator for the age of the deposits:

  - The karst weathering is mostly limited to small pittings and depressions of only millimeter to centimeter depths, which is pointing to a younger subrecent age of only some hundred years;

  - The limited age of the tsunami depositions is documented with the very good preservation of specific supratidal forms (rock pools, small bioerosive notches and edges) on the boulder sides. Evidently, the time period after the deposition was too short to transform the finer surface topography of even exposed boulder sides, indicating an age not older than the Holocene for the events;

  - On the other hand, at some boulders a higher degree of weathering is visible. The different resistance of the interstitial cement and the fossil coral colonies against the weathering process led to a sculpturing with the coral fragments protruding out of the intercrusting limestone, suggesting a relatively longer time of exposure for these boulder depositions.

- All deposits are located in the higher supratidal environment, usually in a vegetation-free belt. In cases where the deposits were transported far inland (>200 m) a scarce shrub vegetation with *Conocarpus* and *Laguncularia* species has been settled. Sometimes wind-shaped trees grow in protected places (e.g. Boka Washikemba, Bonaire). The vegetation in this extreme and dry environment will need several hundred years to reach this stage of succession (minimum age);

- The soil development after the deposition has only reached initial stages in chemical soil formation processes, suggesting a maximum deposition age limited to the Holocene;

- Mining activities transformed the deposits in parts extremely. The beginning of the mining operations of the already existing debris formations reaches back at least 250 years;

- Historical buildings (e.g. fortifications) have not been affected by tsunamis, instead sometimes tsunami boulders have been incorporated in the building, like in the wall of a small ammunition depot in the area of the Sheraton Hotel, Piscaderabaai, on Curaçao. Destructive tsunamis should have occurred only prior to the buildings;

- In the historical record no written or oral sources describing a tsunami impact on the ABC-islands exists. This suggest a time span of minimum 350-400 years without the occurrence of any tsunami event, presumably since the Dutch occupation in 1634.

Summarizing the conclusions deduced from different relative age indicators, we could determine the minimum age for the tsunami debris with high probability to at least some centuries, most likely they are older than the Dutch occupation (1634) or even the occupation by the Spaniards (1527). The maximum age range is limited to the Younger Holocene as evident in particular by chemical and biological weathering processes and the spatial relation of the debris formations to the recent sealevel highstand, which reached the present level between 5000 and 6000 BP in this part of the Caribbean, since when it remains very stable (RULL, 2000).

However, the resolution of relative age dating is insufficient to establish a more detailed chronology of the tsunami impacts, so that absolute age dating with the radiocarbon method was applied to provide further insights into the depositional history.

6.2 Absolute Age Dating

Absolute dating of paleotsunami deposits is rather difficult at a time-scale accuracy of decades. The standard dating technique is radiocarbon, but sampling of appropriate material, i.e. alive during and then killed by the event, as opposed to transport of older material, can be problematic, as so can contamination and correction issues complicate the dating accuracy. Radiocarbon age determinations from 45 samples were performed from different geomorphologic units (boulders, ramparts, ridges) and on different material (vermetids, corals, gastropods). Conventional radiocarbon dating of 43 samples supplied a non-calibrated age range from 370 ± 32 to 4222 ± 49 years BP (Tab. 15). The uncalibrated age data show a clustering in three main time units around

Tab. 15: Conventional radiocarbon ages for 45 samples from Aruba, Curaçao and Bonaire.

Lab.- Nr.	Sample-Nr.	Location	Height asl [m]	distance to coast [m]	sample	debris type	conv. ^{14}C Age BP	δ ^{13}C
				Aruba				
Hd-21530	AUA 1	Cudarebe	5	20	Montastrea ann.	ridge	2217 ± 41	-0.9
Hd-21607	AUA 2	Cudarebe	5	20	Vermetids	ridge	591 ± 45	2.5
Hd-21539	AUA 3	California LH	5	80	Vermetids	boulder	3571 ± 49	3.1
Hd-21546	AUA 4	California LH	4.5	75	Vermetids	boulder	3519 ± 39	3.2
Hd-21563	AUA 5	Druif	7	70	Vermetids	ridge	1752 ± 42	2.8
Hd-21676	AUA 6	Druif	7	70	Acropora palm.	ridge	3349 ± 43	-0.8
Hd-21514	AUA 7	Gabbroformation	4	20	Vermetids	ridge	4222 ± 49	2.4
Hd-21608	AUA 8	Noordkaap	9	32	Cittarium pica	ridge	1410 ± 34	2.9
Hd-21564	AUA 9	Sero Coloradu W	4.5	70	Vermetids	boulder	2233 ± 40	2.4
				Bonaire				
Hd-21570	BON 10A	Boka Chikitu S	6	40	Vermetids	rampart	859 ± 32	2.2
Hd-21517	BON 10B	Spelonk LH	5	80	Diploria strig.	rampart	456 ± 37	-0.3
Hd-21543	BON 11	Spelonk LH	4	70	Colpophyllia nat.	rampart	536 ± 35	-1.2
Hd-21513	BON 12	Spelonk LH	5	80	Cittarium pica	rampart	15880 ± 110	0.7
Hd-21571	BON 13	1km E Willemstoren	1.5	40	Cittarium pica	ridge	619 ± 35	3.2
Hd-21609	BON 14	1km E Willemstoren	2	35	Vermetids	ridge	487 ± 35	1.7
Hd-21651	BON 15	2.5km E Willemstoren	1.5	20	Cittarium pica	ridge	535 ± 47	2.4
Hd-21538	BON 16	2.5km E Willemstoren	1.5	25	Diploria strig.	ridge	518 ± 24	0.4
Hd-21652	BON 17	3km E Willemstoren	0.6	3	Cittarium pica	ridge	912 ± 29	3.8
Hd-21598	BON 18	Bartolbaai	1.5	15	Diploria strich.	rampart	1253 ± 43	-0.1
				Curaçao				
Hd-21493	CUR 21	Jan Thiel Baai	4	15	Acropora palm.	ridge	674 ± 37	0.4
Hd-21615	CUR 22	Jan Thiel Baai	3.5	20	Diploria strig.	ridge	464 ± 32	-1.1
Hd-21616	CUR 23	Slangenbaai W	4	20	Vermetids	boulder	1277 ± 33	1.8
Hd-21507	CUR 24	Slangenbaai W	4	15	Vermetids	boulder	645 ± 28	0.3
Hd-21512	CUR 25	Buoy 1	2.5	3	Diploria strig.	rampart	464 ± 30	0.7
Hd-21621	CUR 26	Boka Grandi E	5	40	Vermetids	rampart	3428 ± 43	-0.1
Hd-21528	CUR 27	Boka Grandi E	5	35	Acropora palm.	rampart	2511 ± 43	-2.4
Hd-21527	CUR 28	Boka Grandi E	5	35	Acropora palm.	rampart	1469 ± 44	-0.8
Hd-21499	CUR 29	Boka Grandi W	5	60	Diploria strig.	rampart	3620 ± 28	1.7
Hd-21508	CUR 30	B. Kortalein/ Plata	7	70	Acropora palm.	ridge	1399 ± 33	-1.0
Hd-21497	CUR 31	B. Kortalein/ Plata	7	70	Acropora palm.	ridge	1557 ± 45	-1.9
Hd-21485	CUR 32	Dos Boka	7	100	Diploria strig.	rampart	1567 ± 38	0.4
Hd-21622	CUR 33	Dos Boka	8	100	Acropora palm.	rampart	3293 ± 37	-0.1
Hd-21480	CUR 34	Christoffel Park	9	120	Diploria strig.	rampart	728 ± 34	-0.4
Hd-21479	CUR 35	Christoffel Park	10	100	Diploria strig.	rampart	1552 ± 40	0.8
Hd-21547	CUR 36	Christoffel Park	12	180	Acropora palm.	rampart	> 42000	-0.4
Hd-21540	CUR 37	Watamula	2	4	Vermetids	boulder	1673 ± 40	2.3
Hd-21597	CUR 38	Watamula	2	4	Vermetids	ridge	1798 ± 35	2.7
Hd-21486	CUR 39	Noordpunt	5	35	Diploria strig.	ridge	1078 ± 42	1.5
Hd-21496	CUR 40	Noordpunt	5	35	Acropora palm.	ridge	1371 ± 44	-0.8
Hd-21708	CUR 41	Oostpunt	2.5	100	Diploria strig.	rampart	370 ± 32	0.5
Hd-21709	CUR 42	Oostpunt	2.5	100	Diploria strig.	rampart	438 ± 25	0.1
Hd-21719	CUR 43	Oostpunt	2.5	80	Acropora palm.	rampart	1557 ± 34	0.5
Hd-21723	CUR 44	Oostpunt	2.5	80	Vermetids	boulder	842 ± 33	3.7
Hd-21696	CUR 45	Awa di Oostpunt	2.5	3	Diploria strig.	ridge	477 ± 30	0.5
Hd-21710	CUR 46	Awa di Oostpunt	2.5	3	Acropora palm.	ridge	2010 ± 38	0.6

500 BP, 1500 BP and 3500 BP with intermediate periods of only infrequent or no age values. A conventional radiocarbon age does not take into account specific differences between the activity of different carbon reservoirs. In order to ascertain the historical ages of samples, it is necessary to provide an age correction. The ocean reservoir stores vastly more carbon than the atmosphere, particularly in deep

Tab. 16: Values for the reservoir age and Delta R for the Caribbean, Florida, Bahamas, Tortugas and Bermuda (Radiocarbon Lab., Queen's University of Belfast, this study).

Lab.- Nr.	Site	Sample Material	Calendar age	^{14}C-age BP	Reservoir age (*)	ΔR (**)	Reference
L-576A	Jamaica, B.W.I.	Liiona pica (gastropod)	1929	423 ± 42	274 ± 42	-34 ± 42	Broecker & Olson, 1961
L-576F	Jamaica, B.W.I.	Liiona pica (gastropod)	1884	425 ± 41	325 ± 41	-44 ± 41	Broecker & Olson, 1961
CAMS-20513	Cariaco Basin, Venezuela	Globigerina bulloides (planktonic foraminifera)	1930	490 ± 40	341 ± 60	32 ± 60	Hughen et al., 1996; pers. comm. 2000
CAMS-20514	Cariaco Basin, Venezuela	Globigerina bulloides (planktonic foraminifera)	1900	460 ± 70	360 ± 50	5 ± 50	Hughen et al., 1996; pers. comm. 2000
HD-21659	Caracasbaai, Curaçao	Acropora palmata (coral)	1920	438 ± 21	314 ± 22	-18 ± 21	this study
HD-21660	Caracasbaai, Curaçao	Acropora cervicornis (coral)	1920	375 ± 25	251 ± 26	-81 ± 25	this study
HD-21673	Caracasbaai, Curaçao	Acropora cervicornis (coral)	1920	586 ± 24	462 ± 25	130 ± 24	this study
HD-21675	Caracasbaai, Curaçao	Acropora cervicornis (coral)	1920	595 ± 24	471 ± 25	139 ± 24	this study
Reginal mean: Caribbean Sea						29 ± 10	
SI-?	Golding Cay, Bahamas	Acropora palmata (coral)	1912	594 ± 66	484 ± 66	144 ± 66	Lighty et al., 1982
SI-?	Tortugas, Florida	Acropora palmata (coral)	1884	584 ± 51	484 ± 51	114 ± 51	Lighty et al., 1982
L-576B	Bahama Islands	Strombus rninor (gastropod)	1950	428 ± 42	234 ± 45	-44 ± 44	Broecker & Olson, 1961
L-576G	Bahama Islands	Strombus rninor (gastropod)	1885	525 ± 59	426 ± 59	56 ± 59	Broecker & Olson, 1961
LJ-series	The Rocks, offshore Florida Keys, USA	Montastrea annularis (coral rings)	1850	518 ± 16	403 ± 17	31 ± 16	Druffel & Linick, 1978; Druffel, 1982
Reginal mean: Bahamas and Florida						36 ± 14	
LJ-series	North Rock, offshore of Bermuda	Diploria strigosa (coral)	1910	399 ± 40	291 ± 40	-50 ± 40	Druffel, 1997
Reginal mean: Bermuda						-50 ± 40	

(*) Reservoir age = measured marine ^{14}C - atmospheric ^{14}C age at time t as defined by Stuiver, Pearson & Braziunas, 1986.
(**) ΔR = measured marine ^{14}C – marine model ^{14}C age at time t (Stuiver et al., 1998)

waters. Depending on the rate of mixing from below, the surface ocean typically has ^{14}C ages 400-1600 years older than the atmosphere, reflecting the marine reservoir age (Hughen et al., 2000; Stuiver & Braziunas, 1993).

Therefore, in any radiocarbon-dated marine record, the magnitude and stability of the reservoir age with time is an important issue. The standard marine calibration routines (MARINE98, CALIB4) incorporate a time-dependent global ocean reservoir correction of about 400 years (Stuiver, Reimer & Braziunas, 1998). That corresponds with the results of Hughen et al. (1996, 2000), who determined the present day Cariaco Basin reservoir age on the basis of two sediment samples of known recent age and averages it with ~420 years. The Carioca Basin is situated geographically close to this study area on the Caribbean coast of Venezuela near Cumana and the islands Tortuga and Isla Magarita. To accommodate local effects, the difference DR in reservoir age of the local region of interest and the model ocean should be determined. The regional reservoir ages and this DR variation for the Intra Americas Sea are comprehend in Table 2.

So far for the Caribbean only four reservoir ages and four DR values were published – the DR value with a regional mean of -18 ± 23 years, which would lead to a regional reservoir effect for the Caribbean of 382 years. For the Bahamas/Florida region the average reservoir age amounts to 436 years taking into account

Tab. 17: Calibrated absolute ages for 43 samples from Aruba, Curaçao and Bonaire, sorted after conventional ^{14}C age (calibrated with MARINE98 and CALIB4 after STUIVER, REIMER & BRAZIUNAS, 1998), Reservoir age = 429 years.

Lab.- Nr.	Sample-Nr.	Location	conv. ^{14}C Age BP	cal. age 1σ	cal. age 2σ
Hd-21514	AUA 7	Gabbroformation	4222 ± 49	cal BC 2407-2262	cal BC2459-2182
Hd-21499	CUR 29	Boka Grandi W	3620 ± 28	cal BC 1572-1492	cal BC1608-1449
Hd-21539	AUA 3	California LH	3571 ± 49	cal BC 1517-1414	cal BC1598-1375
Hd-21546	AUA 4	California LH	3519 ± 39	cal BC 1452-1382	cal BC1505-1342
Hd-21621	CUR 26	Boka Grandi E	3428 ± 43	cal BC 1380-1261	cal BC1414-1207
Hd-21676	AUA 6	Druif	3349 ± 43	cal BC 1284-1155	cal BC1353-1089
Hd-21622	CUR 33	Dos Boka	3293 ± 37	cal BC 1204-1079	cal BC1265-1013
Hd-21528	CUR 27	Boka Grandi E	2511 ± 43	cal BC 224-126	cal BC335-55
Hd-21564	AUA 9	Sero Coloradu W	2233 ± 40	cal AD 96-207	cal AD61-251
Hd-21530	AUA 1	Cudarebe	2217 ± 41	cal AD 122-231	cal AD74-263
Hd-21710	CUR 46	Awa di Oostpunt	2010 ± 38	cal AD 381-449	cal AD328-514
Hd-21597	CUR 38	Watamula	1798 ± 35	cal AD 614-672	cal AD569-695
Hd-21563	AUA 5	Druif	1752 ± 42	cal AD 652-704	cal AD609-746
Hd-21540	CUR 37	Watamula	1673 ± 40	cal AD 698-784	cal AD672-830
Hd-21485	CUR 32	Dos Boka	1567 ± 38	cal AD 806-904	cal AD774-967
Hd-21497	CUR 31	B. Kortalein/ Plata	1557 ± 45	cal AD 810-919	cal AD772-988
Hd-21719	CUR 43	Oostpunt	1557 ± 43	cal AD 822-908	cal AD785-969
Hd-21479	CUR 35	Christoffel Park	1532 ± 40	cal AD 822-919	cal AD781-985
Hd-21527	CUR 28	Boka Grandi E	1469 ± 44	cal AD 912-1020	cal AD879-1044
Hd-21608	AUA 8	Noordkaap	1410 ± 34	cal AD 1001-1046	cal AD969-1073
Hd-21508	CUR 30	B. Kortalein/ Plata	1399 ± 33	cal AD 1010-1051	cal AD983-1083
Hd-21496	CUR 40	Noordpunt	1371 ± 44	cal AD 1023-1090	cal AD988-1169
Hd-21616	CUR 23	Slangenbaai W	1277 ± 33	cal AD 1123-1213	cal AD1064-1251
Hd-21598	BON 18	Bartolbaai	1253 ± 43	cal AD 1155-1242	cal AD1068-1282
Hd-21486	CUR 39	Noordpunt	1078 ± 42	cal AD 1300-1366	cal AD1276-1410
Hd-21652	BON 17	3km E Willemstoren	912 ± 29	cal AD 1434-1466	cal AD1416-1487
Hd-21570	BON 10A	Boka Chikitu S	859 ± 32	cal AD 1457-1502	cal AD1441-1528
Hd-21723	CUR 44	Oostpunt	842 ± 33	cal AD 1467-1517	cal AD1448-1542
Hd-21480	CUR 34	Christoffel Park	728 ± 34	cal AD 1553-1656	cal AD1523-1674
Hd-21493	CUR 21	Jan Thiel Baai	674 ± 37	cal AD 1645-1684	cal AD1582-1703
Hd-21507	CUR 24	Slangenbaai W	645 ± 28	cal AD 1665-1693	cal AD1649-1710
Hd-21571	BON 13	1km E Willemstoren	619 ± 35	cal AD 1675-1713	cal AD1656-1815
Hd-21607	AUA 2	Cudarebe	591 ± 45	cal AD 1683-1816	cal AD1661-1885
Hd-21543	BON 11	Spelonk LH	536 ± 35	cal AD 1722-1883	cal AD1696-1950
Hd-21651	BON 15	2.5km E Willemstoren	535 ± 47	cal AD 1712-1897	cal AD1686-1950
Hd-21538	BON 16	2.5km E Willemstoren	518 ± 24	cal AD 1816-1892	cal AD1727-1950
Hd-21609	BON 14	1km E Willemstoren	487 ± 35	cal AD 1834-1950	cal AD1806-1950
Hd-21696	CUR 45	Awa di Oostpunt	477 ± 30	cal AD 1825-1955	cal AD1805-1955
Hd-21615	CUR 22	Jan Thiel Baai	464 ± 32	cal AD 1835-1955	cal AD1810-1955
Hd-21512	CUR 25	Buoy1	464 ± 30	cal AD 1840-1955	cal AD1815-1955
Hd-21517	BON 10B	Spelonk LH	456 ± 37	cal AD 1840-1955	cal AD1810-1955

the local DR variation of +36 ± 14 years, whereas data from Bermuda yield a local DR variation of -50 ± 40 years. Some pitfalls of local reservoir effect corrections are discussed in detail by SPENNEMANN & HEAD (1996). We tried to verify and precise the reservoir age and DR variation for our study area with the dating of four coral samples (*Acropora palmata* and *A. cervicornis*) of the VAN DER HORST Collection (collec-ted in 1920), which is located at the Zoological Museum of Amsterdam University (Tab. 16). Since the marine model ^{14}C age varies in the time period around 1920 relatively extreme, we decided to take a 5 point moving average of 456 ± 5 (1900 - 1940) as the marine model ^{14}C age at time t for the four samples of Curaçao (pers. comm. B. Kromer, 2001). The resulting regional DR variation for the Caribbean as

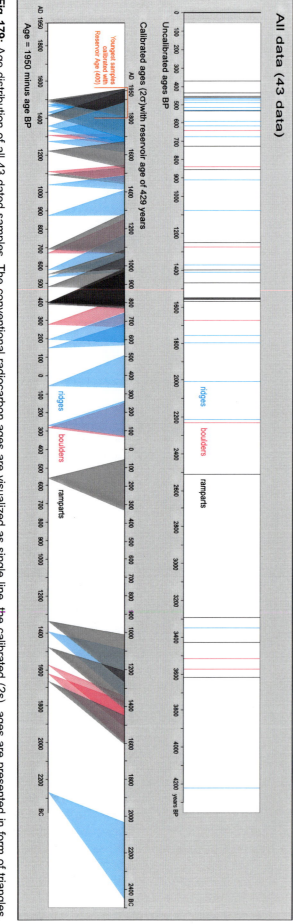

Fig. 179: Age distribution of all 43 dated samples. The conventional radiocarbon ages are visualized as single line, the calibrated (2s) ages are presented in form of triangles. Reservoir age = 429 years.

the weighted mean of the DR values, where the weighting factor is $1/s^2$ (BEVINGTON, 1969), aggregates to +29 ± 10 radiocarbon years, which was considered in the calibration of the conventional radiocarbon ages. The calibrated data set is comprehended in Table 17.

The absolute age distribution shows a concentration in clusters around 500, 1500 and 3500 BP, which extend over a 200-300 time period in each cluster. These results are corresponding well with the hitherto published absolute ages for assumed paleotsunami events in the Intra Americas Sea. SCHUBERT (1994) determined the age of the paleotsunami event in Cumana, Venezuela with 1300 ± 160 BP by $^{230}Th/^{234}U$ ages. Simultaneously radiocarbon datings from three fragments of the same deposit yielded to absolute ages of 470 ± 60, 350 ± 60 and 310 ± 100, respectively. The author suggests that the uranium-series ages are more reliable as a possible ^{14}C contamination might have occurred. JONES & HUNTER (1992) suggested a paleotsunami event approximately 330 years ago at 660 ± 50 BP on Grand Cayman and calculated a calibrated age of AD 1662 (cal. age 1s: 1625-1688) for the boulder accumulations. WEISS (1979) reported radiocarbon dates of 500-700 BP for a possible tsunami in Venezuela related to the disruption of the morphology and sedimentary sequence of the island lagoon at Cayo Sal. The data corresponds with our results for the youngest and the middle cluster. The evidence of a tsunami impact around 3500 BP extends the paleotsunami record for the Caribbean for another 2000 years.

A certain age deviation of the absolute values in our data set is not surprisingly as they are to be expected by e.g. dating of different species, sampling different growth sections (outer or inner growth bands), by possible contamination or slightly different sample preparation in the laboratory. The youngest cluster might suggest a subdivision of the data in two single units from which two tsunami events within the last 700 years could be deduced, but we regard this rather as an over-interpretation of the method and the dated material. Assuming that the estimation of a local reservoir age of 429 years is correct – as all published data strongly suggest – this would lead to a historical time span considering the sigma 2 values for the date of the event for the oldest tsunami between 1640-1000 BC (Fig. 179). The middle event would have occurred with a 95% probability between 600-1290 AD. The youngest cluster suggest an event between 1400 and the end of 1800 AD. For the interpretation of the data it is essential to point out that the absolute age values represent maximum ages for the dislocation event. Therefore, the most probable age of the events might be slightly shifted to the younger margin of the clusters. The deductions from the relative age indications rather point to an older age for the youngest event.

Considering these findings the standard reservoir age of 400 years and a local reservoir effect correction of +29 years might be questioned at least for this part of the Caribbean basin, but a detailed discussion of the subject is far behind the scope of this study.

The distribution of the age values supports the interpretation of the coarse debris deposits as tsunamigen, and is inconsistent with a hypothesis of a storm-induced origin. If storm events would have contributed at least partly to the depositions, we could expect an even distribution of the data samples over the time period, when the Holocene sealevel reached more or less the present height around 5000 BP (RULL, 2000) (see Fig. 15).

In contrast, data from the first 1000 years of high Holocene sealevel are lacking completely, as so from 4200 to 3600 BP, 3300 to 2500 BP, 2100 to 1800 BP or the past 350 years BP. We suggest – considering that once every 100 years the islands experience considerable damage by tropical cyclones, and on average once every 4 years a tropical cyclone occurs within a radius of approx. 100 nm – that even extreme storms have not transported coarse sediments in a significant amount far from the supratidal environment. The distribution of the age data rather corresponds with the field observations in that the zone potentially reached by storm wave action along the most exposed coasts is entirely bare of sediments, even though the rough surface in the rock pool zone represents excellent sediment traps. It is concluded that the clustered data represents paleotsunami events. Scattered data in between the clusters can be explained by the incorporation of older fragments in the deposits, which have stayed over a longer time period in the foreshore zone and were picked up and transported onshore during the following tsunami event.

Only two samples (BON 12, CUR 36) were dated significantly older. The sample CUR 36 was taken from a coral fragment of *Acropora palmata* and dated to >42,000 BP, which lay beyond the limit of this method. Presumably, the fragment represents an extremely well preserved piece of coral from the Pleistocene coral reef terrace. For the sample BON 12, a rounded vermetid rim, an age of 15,800 BP was calculated. We regard this as a reliable age, which might be interpreted as follows:

- The fragment was broken off from a bench in the intertidal zone of the sealevel of that time, which was at least 80 m lower than today, and transported by one of the Holocene tsunamis onshore (it is deposited at a location of the most energetic tsunami impact of all the islands);

- It was transported by successive transgression of the sealevel during lateglacial and postglacial times and incorporated in young Holocene sediments, from which it has been picked up by a tsunami;

- The fragment was dislocated from that depths by sand mining with offshore platforms, as it occurred occasionally in the past.

The data distribution will further be discussed concerning the three different debris formations, the type of the dated material as well as for each island separately. Finally, open questions will highlight the needs for further age determinations.

6.2.1 Biological Differences of the Age Distribution – the Species

A similar picture is obtained by comparing the age distribution with regard to the different main species. Absolute ages were determined from two coral species (23 ages), from one mollusks (4 ages) and one vermetid species (14 ages) (Fig. 180).

The selection of corals was restricted (with two exceptions) to the species *Acropora palmata* and *Diploria strigosa*. The elkhorn-like species *Acropora palmata* forms large colonies, commonly 4 m across, 2 m high with bases 0.4 m thick with growth rates of up to 9 cm/yr (VERON, 2000). Their habitat is the shallow outer reef slope exposed to wave action. Along the leeward coasts of Curaçao and Bonaire the depth range extends from 0.5-6 m depending on the wave energy level (VAN DUYL, 1985). Between 0.5 and 2 m depth in the upper terrace zone the community was found most frequently, mainly as an intermittent belt along rubble beaches (VAN DUYL, 1985).The massive head coral *Diploria strigosa* is common in most reef environments, especially on shallow slopes and lagoons, and according to VAN DUYL (1985) on Curaçao and Bonaire the head coral group can be found in all reef zones and all wave energy environments. The growth rate is with 2mm/yr considerably slower, a 1-2 m high colony represents therefore a 200 to 400 year growth history. The gastropod *Cittarium pica* prefers subtidal rock habitats where it inhabits waters too rough for other grazers. *Spiroglyphus irregularis*, a vermetid, is found primarily in the intertidal zone, where it forms thick bioconstructive accretions on the fossil limestones.

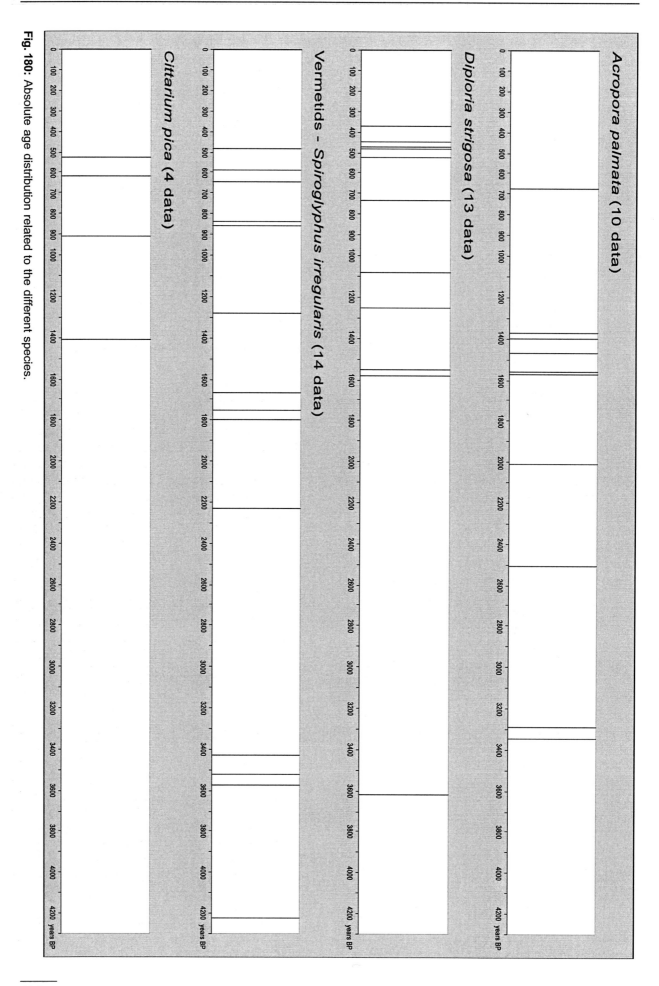

Fig. 180: Absolute age distribution related to the different species.

With the aim to define the age of breaking-off by the tsunami events as accurate as possible, the material in case of the coral samples was taken from fresh-looking, not rounded fragments and there out of the youngest growth zones, e.g. the tips of *Acropora palmata* branches. The dating of vermetids include some pitfalls, which have to kept in mind with the interpretation of the absolute ages. Normally, the accretions are not preserved as complete as they were present at the time of the destructional event and only attached remnants will be found. In so far it is unclear how far the sampling point was located from the living outer growth belt during the time of the destruction (see KELLETAT & SCHELLMANN, 2001a). It might be that the vermetid pebble dated to 4222 ± 49 represents a fragment of the inner and therefore older layer of a vermetid accretion. Nevertheless, the two absolute ages of 3571 ± 49 and 3519 ± 39 are regarded as highly reliable.

The samples were derived from an accumulation of an extraordinary well-preserved field of vermetid boulders at California Lighthouse (Aruba), which could not be observed at any other locations on the ABC-islands. The absolute ages of *Diploria* coral fragments tend to concentrate around younger ages, in particular when they are incorporated in ridges (CUR 22, 45, BON 16), which are located close to the present shoreline or along coastal stretches where the height of the Lower Terrace is extremely low, as the case for CUR 25 close to Piscaderabaai or CUR 41 and CUR 42 at Oostpunt. The calibrated ages (2s) suggest a rather recent deposition of the samples between the time period 1705-1955. This samples might be as well linked to storm deposits, which are occasionally mixed into the tsunami deposits, especially when the samples were taken at the upper surface or the seaward margin of the deposits. The size of the sampled *Diploria* heads usually did not exceed 30-40 cm in diameter, and regarding the size of the fragments a deposition by hurricanes cannot be excluded as was proven by hurricane *Lenny* at Slagbaai, Bonaire, where coral fragments of similar size and shape have been deposited up to 50 m inland in a distinctive semicircle pattern (see Fig. 146). Nevertheless, *Diploria* species are well suitable for absolute dating as obvious by the reliable data values for the older time range, where they reflect mainly ages for rampart formations.

It may be concluded that in order to avoid a mistake between a hurricane or tsunami deposition at locations close to the shoreline, the sampling should be restricted to coarser fragments, like *Acropora* branches. The sampling of *Acropora* species has the additional advantage, that the yearly growth bands are much wider than those of *Diploria* species, which ascertain that with a sampling size of ~5 × 5 cm the skeleton material originates from only very few years. Therefore, the absolute age determination is more accurate, than absolute ages obtained from sample material, which represents due to slower growth rates a longer time interval like e.g. *Diploria* species. But generally, the obtained absolute data spread over the whole range of the Younger Holocene and no species-dependant age relation can be observed within the data set.

6.2.2 Geomorphological Differences of the Age Distribution – the Debris Formations

The data set contains of 18 samples collected out of rampart formations, 18 samples out of ridge formations and only few (7) samples could be taken at single boulders (Fig. 181).

Overall, the collection of appropriate sampling material within the rampart formations and for the boulder formations is difficult. Both formations originate mainly out of the supratidal and consist of Pleistocene limestone. Holocene material is only occasionally attached in form of vermetid accretions on boulders, and unfortunately no boring mollusks (e.g. *Lithophaga* sp.), generally very suitable for absolute age dating, were found. For the dating of the rampart formations, we were therefore restricted to date incorporated corals, which only represent a small amount of all fragments. But still the data distribution of the absolute ages clearly gives evidence of the three age clusters around 350-550 BP, 1400-1600 BP and 3300-3600 BP. The young absolute ages (5) were derived from the ramparts at Spelonk Lighthouse (2), Oostpunt (2) and close to Piscaderabaai (1). They cover a time period from 370-536 BP, which would lead to calibrated historical dates from 1685 to 1955. That historical date clearly contradicts to the interpretation of the relative age indications and it is highly unlikely that such a recent event would not be mentioned in any historical record. On the other side, a storm-induced deposition can be excluded with regard of the depositional distance to the coastline. At Spelonk Lighthouse, Bonaire, the sampled fragments have been deposited in 70-80 m distance to the coastline, at Oostpunt, Curaçao, even in a distance of 100 m. Here, a denser sampling net for absolute age determinations would be necessary to clarify the present results. The sample at Buoy 1, close to Piscaderabaai, is deposited at a height of 2.5 m in a distance of 3 m from the shoreline and it might represent a recent storm deposit as well.

The few absolute ages obtained from boulders are relatively evenly scattered over the entire time period. Four absolute ages were obtained from measured

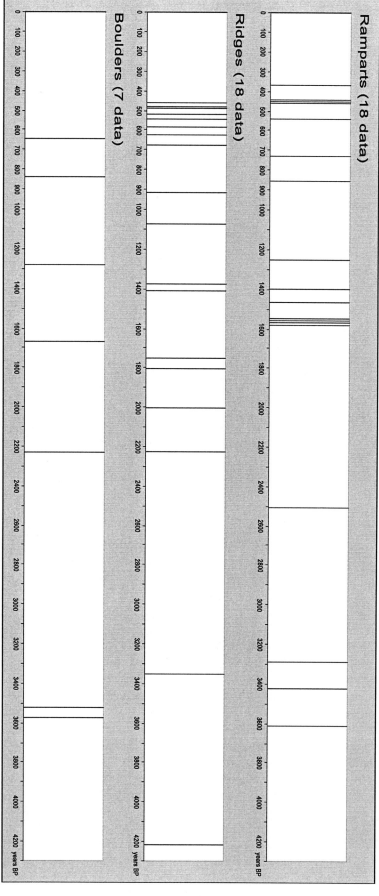

Fig. 181: Distribution of absolute ages in relation to the different debris formations.

boulders, for which the hydrodynamic formulas of the required wave height were applied (Table 18).

The wave height represent only minimum wave heights as with the applied hydrodynamic formulas only the moment of overturning is calculated and factors as e.g. break-off forces, the transport against gravity or measured average b-axis of boulder versus real b-axis are not considered in that calculations. From this comparison it might be carefully deduced that the tsunami event around 3500 BP documented with the two vermetid boulders at California Lighthouse on Aruba was of higher magnitude than the youngest event. In contrast, with regard to the far-reaching inland extent of the young rampart formations at Spelonk and Oostpunt, which are associated with impressive boulder assemblages, it seems likely, that the youngest impact reached the higher magnitude. Although the field observations suggest a contemporaneous deposition of the ramparts and the boulder assemblages, an older depositional age of the boulders equivalent with a severe tsunami impact cannot be excluded, as in particular on Bonaire no boulders could have been dated directly. On the other side, the absolute age calculated for the boulder located at Oostpunt confirms a young deposition age.

The absolute data of the ridges show a concentration between 460 and 670 BP, but than only infrequent data values are scattered between 900 and 2200 BP, here a clustering is not perceivable. The oldest tsunami event (3500 BP) is documented with only one absolute age of an *Acropora* fragment from the ridge at the location Druif in Aruba. As most of the samples were taken from the surface of the ridges, the older events might be hidden in the basis layer of the ridges. To clarify this suggestion possible future dating approaches should focus on vertical dating profiles through ridge deposits.

Tab. 18: Absolute ages and required minimum depositional wave heights for four measured boulders (for formulas see Chapter 3).

Lab.-Nr.	Location	Boulder size [m]	conv. ^{14}C age BP	cal. age 2σ	Storm wave [m]	Tsunami wave [m]
Hd-21539	California LH, Aruba	4.7 × 1.4 × 0.6	3571 ± 49	BC1598-1375	30	11
Hd-21546	California LH, Aruba	2.8 × 1.6 × 0.8	3519 ± 39	BC1565-1342	27	10
Hd-21616	Slangenbaai, Curaçao	2.5 × 1.8 × 1.6	645 ± 28	AD1064-1251	18	6
Hd-21723	Oostpunt, Curaçao	5.0 × 1.2 × 1.0	842 ± 33	AD1448-1542	14	5

6.2.3 Geographical Differences of the Age Distribution – Comparison of the Islands

Aruba

The data from Aruba (Fig. 182) show one absolute age at 591 BP, two around 1500 BP and three close to 3500 BP, which are the most probably ages of three young Holocene tsunamis. The oldest and the middle tsunami event are well documented along the windward coast in the extensive ridge formation at Druif and with the boulder deposits at California Lighthouse (Fig. 183). At Druif ridge, the oldest absolute age (4222 ± 49 BP) of the data set was obtained on a vermetid pebble. At the location Noordkap, the absolute age of a gastropod sample within a ridge deposit at a height of +9 m asl confirms a high magnitude depositional event around 1500 BP. The only samples (AUA1, AUA9) with ages around 2200 BP are located at the northern and southern tip of the island.

Bonaire

On Bonaire, eight of the nine absolute ages are documenting exclusively the youngest tsunami impact (see Fig. 182). Two samples have been taken at the location Spelonklighthouse, the area with the most impressive boulder assemblages (size, amount, distance to coastline) suggesting the strongest energetic tsunami impact of all the islands (Fig. 184). Unfortunately, no samples were found on the boulders themselves. Both samples were collected out of the associated rampart formation 70-80 m distant to the coastline and gave an age of 456 ± 37 BP (BON 10B) and 536 ± 35 BP (BON 11), respectively. The absolute age of a vermetid encrustion attached to a smaller boulder at Boka Chikitu, some kilometers south of Spelonk, gives a slightly older value of 859 ± 32 BP. The calibrated time span of this sample is calculated with 1441-1528 AD, which seems to be congruent with the relative age estimation.

Whereas the calibration of the two younger samples, in particular BON 10B with an calibrated age of 1810-1955 AD, suggest a very recent impact, which is unlikely as a tsunami impact of this magnitude supposedly would be documented in historical records. In order that the adjacent areas are heavily mined, a disturbance of the ramparts might be possible. The ridges in the southern part of Bonaire show similar young absolute ages, ranging from 487 ± 35 to 619 ± 35 BP for samples collected at the surface of the ridges. A gastropod sample taken 0.5 m below the surface of the ridge was dated to 912 ± 29 BP, which correlates to a calibrated age of 1416-1487 AD. The calibrated ages of the surface samples point to a depositional event between the early 17th century to the beginning of the last century. These calibrated ages seems strikingly as a major storm or hurricane event should be documented in historical documents as the ridges are located in the area of the economically important salt pans, which were exploited since 1634 by the Dutch.

For the leeward coast only one absolute age from the rampart deposit of Boka Bartol was acquired, which hints with an absolute age of 1253 ± 43 BP (calibrated: 1068-1282 AD) to the middle tsunami event, but here a denser sampling network is necessary to confirm this result.

Curaçao

The data from Curaçao (25) reflect all three tsunami events (see Fig. 182), but a remarkable geographical distribution can be observed as well (Fig. 185).

The most samples with a young age (8 from 9) are incorporated in the deposits of the leeward coast and the Oostpunt area, whereas in contrast, all samples in the northern part of Curaçao reflect the middle and the oldest tsunami event relatively congruent. Only one sample with a young age (728 ± 34 BP) can be found in the northern part of the island within the rampart at Boka Grandi. In particular, the extensive rampart formations on the windward northeastern coast reflect the middle and older tsunami event apparently. At Oostpunt, a younger and the middle event are represented in the debris formations. At Slangenbaai, two absolute ages obtained from vermetid accretions on boulders reflect a high magnitude event at 645 ± 28 BP (cal: 1649-1710 AD) and at 1277 ± 33 BP (cal: 1064-1251 AD), correlating with the younger and middle tsunami event. The ridge deposits of Curaçao show mostly similar young ages like the ridges of Bonaire.

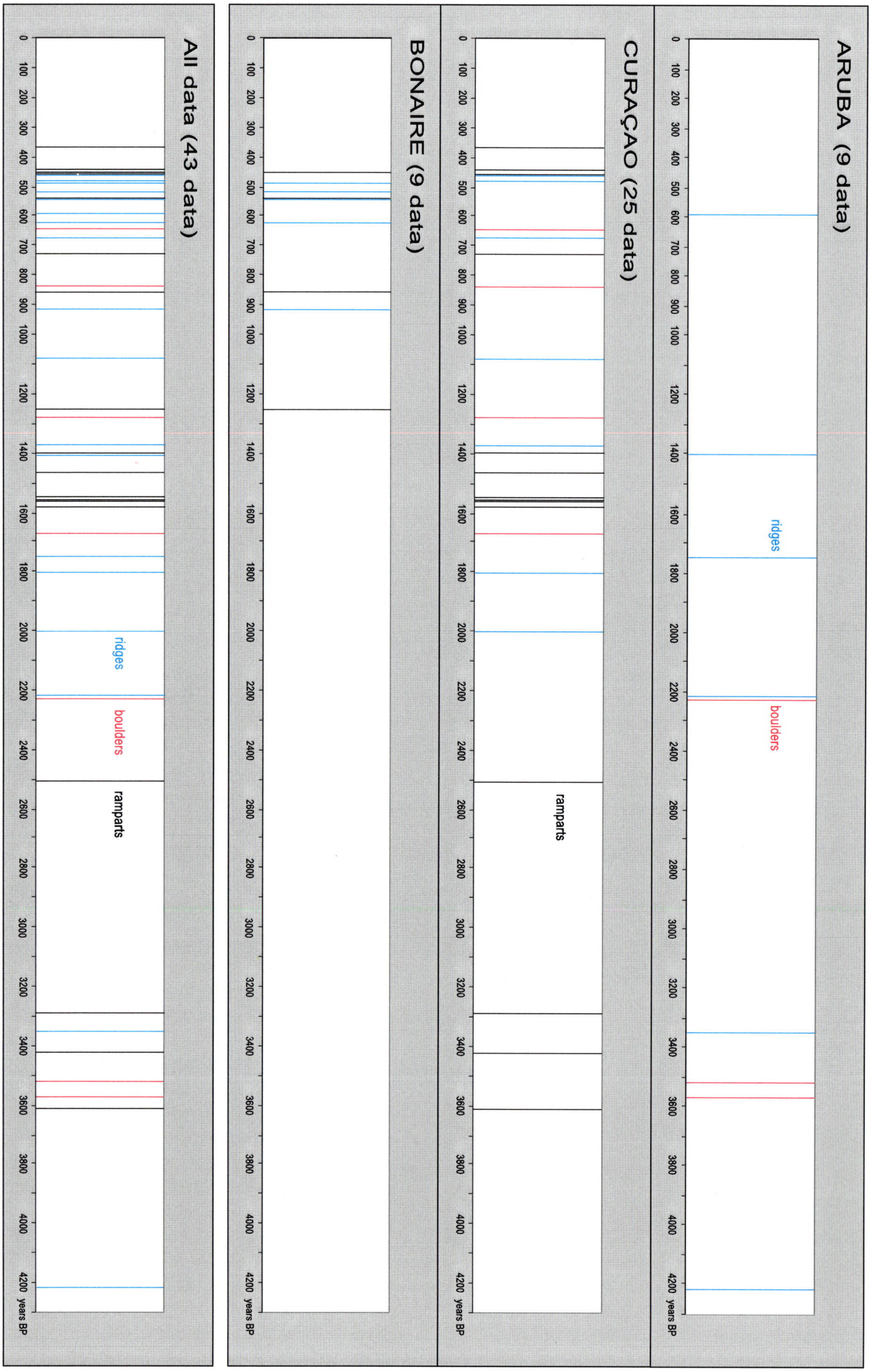

PALEOTSUNAMIS IN THE CARIBBEAN

Fig. 182: Comparison of the absolute radiocarbon ages of all three islands.

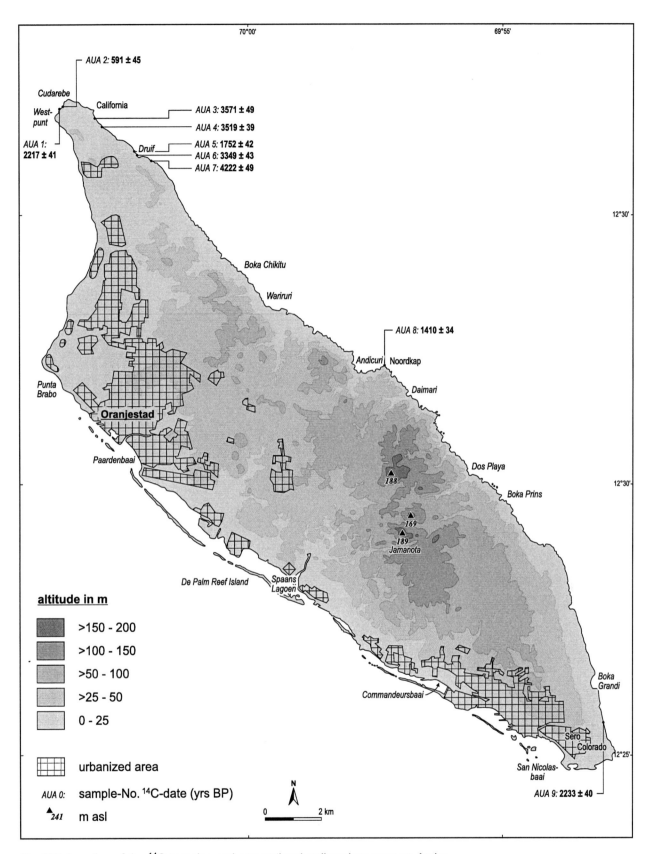

Fig. 183: Location of the ¹⁴C-samples and conventional radiocarbon ages on Aruba.

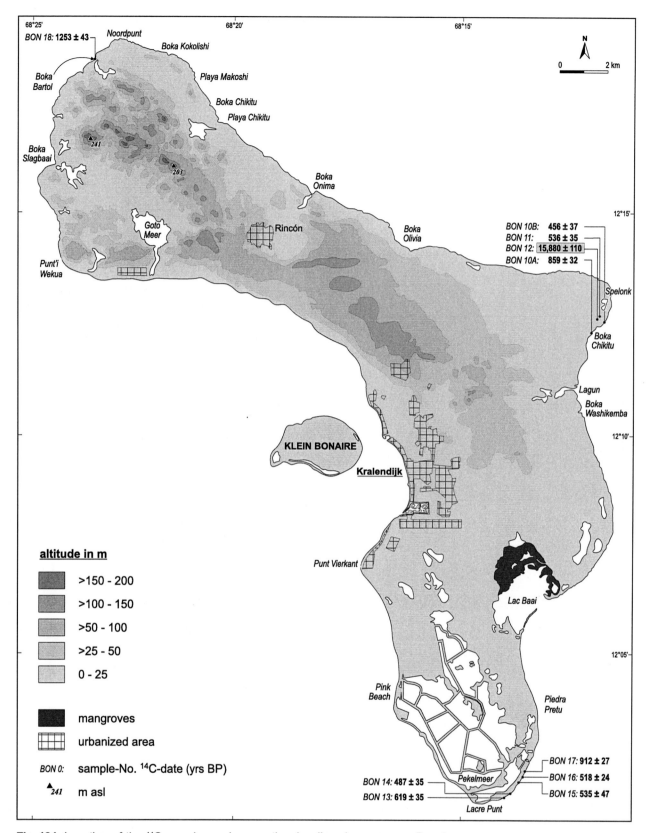

Fig. 184: Location of the ^{14}C-samples and conventional radiocarbon ages on Bonaire.

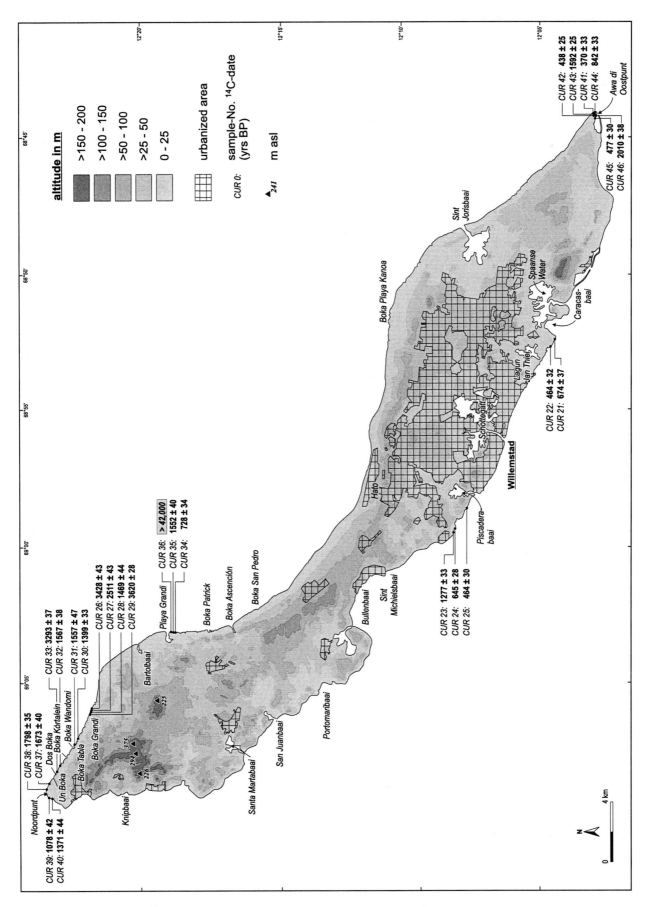

Fig. 185: Location of the ^{14}C-samples and conventional radiocarbon ages on Curaçao.

6.2.4 Comparison of Relative and Absolute Ages

Geomorphology and surface topography at some locations permit to distinguish between a seaward and a landward subunit of some ridge or rampart formations, but a reliable age determination on the basis of these evidences is unfeasible, because all fragments have been weathered beyond a certain intensity. The status of weathering does not reflect age differences of 1000 or 2000 years and is therefore not suitable to differentiate into the three clusters, which could be determined by the absolute ages.

From the field evidences we can only reliably derive the conclusion, that the time period after the last tsunami was sufficiently long enough, as all material is significantly weathered and biological and chemical weathering led to the destruction of the tsunami imprints in the supratidal rock pool area and the cliff front. It is most likely that the youngest tsunami impact is older than any human evidence in the possible reach of a tsunami wave. For the youngest event, the calibrated absolute ages conflict partly with the evidences derived from the relative age indications and the documented historical record. Some of the young absolute data can be explained by the possible dating of a storm deposit, which is intermixed with the tsunami deposits.

Nevertheless, a very recent event in the 17th century is congruent with a suggested paleotsunami event at AD 1662 in Grand Cayman. But for a better understanding more detailed absolute age determinations are required. However, for the establishment of a higher resolution tsunami chronology the application of absolute age dating methods is essential. The relative and absolute ages presented in this study should be regarded as a first approximation and constitute a basis for further research. We were restricted to date only the most significant deposits, and a future intensified dating project should establish a more objective network of samples. In this study the main percentage of samples were derived from Curaçao, in order to get a more accurate picture more samples should be taken from Aruba and Bonaire. Overall, some questions and problems remain unanswered, but a more detailed age determination is out of the scope of this project. We hope that nevertheless this limited insight of the tsunami history of the Leeward Lesser Antilles will contribute to a better knowledge of the tsunami chronology on a Caribbean-wide scale.

6.2.5 Future Applications

Besides increasing the amount of absolute ages, further dating approaches might focus on:

- Transects over the well-developed tsunami ridge south of Bonaire in order to enhance information on stratigraphy. It cannot be excluded that this ridge contains more than only the youngest tsunami event;

- A transect over the double ridge in southern Curaçao with a similar aim. But both transects will require movements of about 1000 tons of coarse material;

- Quantifying the status of weathering and karstification on dislocated debris fragments and boulders by systematic micro-measurements with regard to different species or height and distance from the surf zone;

- Determining a more exact age of the vegetation in and around the different formations (e.g. at Boka Grandi, Jan Thiel and Oostpunt on Curaçao; Spelonk on Bonaire);

- Long-term micro-erosion meter measurements in the supratidal rock pool zone to acquire data about the development and recovering time of these features;

- Determining the absolute age and development of benches in different exposures to recognize events of destruction by tsunamis;

- Establish a stratigraphy for fine sediments in embayments of the sheltered and exposed sides of the islands with borings.

7. Linking of the Paleotsunami Deposits to Source Mechanisms and Runup Heights

In the case of paleotsunamis it is extremely difficult to attribute individual tsunami deposits to a certain source mechanism (DAWSON, 1999). Additionally using paleotsunami deposits to calculate runup heights in order to produce inundation maps will represent at best only minimum values. For example, it is very often unclear whether the tsunami event has been triggered solely by offshore earthquakes, submarine sediment slides or by a combination of these processes. Unfortunately, most of the deposits cannot provide direct information about the source mechanism of the tsunami. In so far the linking of the paleotsunami to a certain source mechanisms remains mostly tentative. SCHUBERT (1994) associates the paleotsunami evidence on the Caribbean coast of Venezuela – located geographically close to the ABC-islands – due to their limited extend most plausibly with the occurrence of a local submarine landslide related to a prehistoric large-magnitude earthquake. Nevertheless, he points out that along the whole coastline of the Venezuelan Coast Range of the Caribbean Mountains, there is ample evidence of landslides as well as historical evidence of the occurrence of strong earthquakes with maximum estimated surficial magnitudes (M_s) of up to 8.5 (SCHUBERT, 1994).

The spatial distribution of the debris deposits of the Leeward Netherlands Antilles extent over a region of more than 200 km (distance Aruba-Bonaire), which clearly demonstrate tsunami events of at least regional scale. The results of the absolute age determination have proven that two of the three assumed impacts (the youngest and middle event) took place on all three islands – only the oldest event is lacking in the present data set for Bonaire. The spatial distribution of the debris deposits show on all islands a clear dominance along the windward northeast exposed coastlines. The generating mechanisms of paleotsunamis of the described magnitude is unknown, but most likely they are related to seismic activity in the northeastern part (0-90° sector) of the Caribbean along the faults of the Caribbean Plate boundaries (Fig. 186). Another potential source region is the Southern Caribbean Plate Boundary Zone along the northern Venezuelan continental margin with clear evidence of neotectonic right-lateral strike-slip deformation including uplift and subsidence of large fault blocks along the fault zones. In particular the region in the vicinity of Cumaná is known for the occurrence of paleotsunami events (see Tab. 10). Additionally, these tectonic settings together with

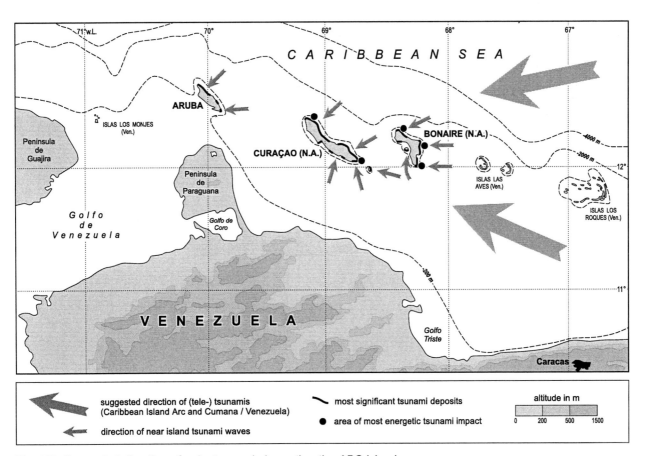

Fig. 186: Suggested direction of paleotsunamis impacting the ABC-islands.

the occurrence of steep subsea slopes might favor the origination of submarine slides, which could have triggered the paleotsunamis.

Even if the source mechanisms themselves remain unclear, the direction of the origin of the tsunami could be derived relatively from:

- Comparison of the intensity and magnitude of the tsunami impact along the coastlines of the ABC-islands through the inland extension of the deposits, boulder size and ridge/rampart development;
- Differentiation of the deposits with regard to the exposition of the coastlines;
- Analysis of the orientation of the fragments with situmetric measurements and via the transport direction of single boulders, but here the local coastal configuration and wave refraction can modify the general findings.

Several studies of coastal sediments deposited by paleotsunamis have shown that the highest of these sediment accumulations always occur below the upper limit of tsunami runup (DAWSON, 1999). Therefore, most paleotsunami deposits cannot provide useful indications of former runup limits. Nevertheless, KELLETAT & SCHELLMANN (2001a, b) could derive the relative accurate runup heights of the Cyprus tsunami from the evidence of spatial soil erosion and the destructive impact on the vegetation. For the ABC-islands this method cannot be applied as the soil formations of the Lower Terrace – as the depositional basis for the tsunami debris – show only initial soil development. Mostly the bare karstified surface is exposed and the successional stage of the vegetation is restricted to small shrubs and cacti vegetation due to the climatic conditions and intensive degradation by cattle. However, the estimation of the minimum runup heights can therefore only be derived from the inland margin of the debris extent.

Mostly the inland extent is limited to a distance between 50-150 m, but in cases (e.g. Spelonk, Bonaire) the debris can reach as far as 400 m inland. To the inland margin the debris accumulations usually gets thinner, the deposition more discontinuous and the size of the fragments and boulders decreases. The rough karstified surface might have decelerated the transport capability of the tsunami waves and the structural texture of the limestones could have supported the drainage of the water masses.

Generally, the topography of the Lower Terrace is flat and most probably large parts of the terrace were covered by the waves, particular in the south of Bonaire, Klein-Bonaire and Klein-Curaçao. Here, the water masses most likely have inundated the entire islands and a geomorphologic effective backwash flow with fluvial erosion and sediment re-deposition did not occur. The nature and the scale of back wash sediment transport or reworking is greatly influenced by the nature of the local coastal topography. As stated the Lower Terrace is characterized by a flat and almost horizontal topography, occasionally a depression – the former reef lagoon – is developed.

Consequently, a runup in its literally sense did not take place and rather an unhindered inland inundation could occur without, in turn, strong currents flowing from landward to seaward. The field observations confirm this as in no case we found evidence for geomorphologic effective back wash.

8. Conclusions

The littoral debris accumulations of the Leeward Lesser Antilles and Aruba contain three main geomorphologic distinct types of paleotsunami debris formations, which have been distinguished as boulder assemblages, rampart formations and ridge formations. Hitherto, these deposits represent the most extensive geomorphological evidence for the occurrence of paleotsunamis on a Caribbean-wide scale and so far the islands Aruba, Curaçao and Bonaire were unknown for tsunamis affecting their coastlines. Predominantly, the debris deposits have been accumulated on the northeastern sides of the islands, reaching from sealevel to a height of +12 asl and extending up to 400 m inland. On a regional scale, the extent and amount of tsunami debris diminishes from east to west with the highest energy impact on Bonaire in the east and a considerable lower impact on Aruba, the most westerly island. Each formation exhibits a distinct morphology and geographic distribution related to a certain coastal configuration.

Boulder assemblages contain blocks of >100 m³ in volume and with a weight of up to 281 tons. They occur on all islands with the most impressive evidences on Bonaire and Curaçao, but in general, they are coinciding remarkably often with coastal sections, where a bench is lacking, cliff front is nearly perpendicular and the supratidal zone is rather narrow. If the coastal physiography leads to the development of a rather broad supratidal with a more convex cliff profile, the amount of debris increase significant as more material can be derived by the tsunami. That coastal environment favors the development of rampart formations. They occur likewise on all islands with the most developed ones in northeastern Curaçao and along the east-exposed coastal stretch from Spelonk Lighthouse to Lac Baai on Bonaire.

The ramparts are located with their seaward margin in distance of at least 40-50 m from the active shoreline, in cases up to 100 m, at elevations usually ranging from +6.0 to +10.0 m asl, and usually they are becoming more scattered and thin out with increased inland extent. They consist of small to medium sized fragments and show a thickness of some cm to several decimeters with a planar gently land inwards sloping profile. Unfortunately, most of the rampart formations are massively disturbed or even completely removed due to intensive mining exploitations in the past.

The ridge deposits have been differentiated in recent or subrecent debris ridges with direct contact to the coastline and surf zone. These type of ridges dominate and only one observed subrecent debris ridge at Watamula, Curaçao, is located several meters above sealevel and in a certain distance to the coastline. The recent ridge deposits are storm-induced and mainly attributed to hurricane *Lenny* (1999), an extraordinary event due to its west-east tracking route through the Caribbean. They show an asymmetric profile with grain size distributions ranging from centimeters to some decimeters with more gentle seaward slopes and steep inward faces. Smaller fragments of *Acropora cervicornis* are the most common components. The typical subrecent debris ridges are of tsunami-origin and consist of mostly well-rounded platy and rod-shaped coral fragments with some rare limestone blocks present. Predominantly, the rounded material is derived from debris deposits out of the subtidal environment. These ridges occur along the southern, southeastern and western leeward coastlines, where they may extend over several hundred meters with width from 10-50 m and relative heights from 1-3 m.

Aruba, Curaçao and Bonaire have experienced multiple episodes of tsunami impacts that have taken place during the Younger Holocene. Geomorphologic relationships between the debris formations and coastal features, e.g. rockpools and bench development, illustrate that at least a time period of some hundred years since the youngest tsunami event must have expired. Furthermore, the field observations suggest at least two depositional events, but it remained unclear whether as well two tsunami wave trains might be responsible for the morphologic differentiation.

Absolute age determinations with the Radiocarbon method confirm that suggestion by documenting the occurrence of three tsunami impacts during the Younger Holocene around 500 BP, 1500 BP and 3500 BP. Both field observations and relative/absolute age dating indicate that:

1. A storm or hurricane-induced deposition can be definitely excluded and therefore the debris formations could be unambiguously related to tsunami events. In addition, the application of hydrodynamic calculations verifies this suggestion;

2. The tsunami events affected all three islands, and were therefore of at least regional scale. Only on Bonaire, no evidence for the oldest event could be proven with absolute dating;

3. The strongest impacts can be observed along the windward coastlines and to a minor extent along the southern parts of the leeward coastlines;

4. The initial formation and the extent of the debris formations is strongly controlled by the pre-existing coastal configuration and the development of the supratidal belt and the offshore reef;

5. The tsunamis approached the islands from a northeasterly direction, that suggest the generating mechanisms located within the Lesser Antilles island arc and the subduction zones around the Caribbean Plate boundaries;

6. After the occurrence of the youngest tsunami the accumulations have not been transformed by any storm events, beside some seaward margins of ridge deposits, which are in direct contact to the present surfzone and are therefore partly reworked. Subsequent disturbances of the deposits are caused mainly by extensive mining exploitations.

From this new, but still limited knowledge of the occurrence of paleotsunamis with severe magnitudes in the Southern Caribbean, we can derive that potentially catastrophic tsunamis may represent a much higher risk than at present recognized by the governmental organizations and the inhabitants of the Caribbean islands. The risk of severe tsunamis anywhere around the Caribbean is still largely unknown as geomorphologic observations of tsunami evidences are yet very rare and many presumed imprints of tsunamis have not yet been found, studied and mapped in appropriate detail. From what we know at present, the Intra-Americas Sea Region has averaged one damaging tsunami every 26 years, but since they have not experienced one in 53 years, a destructive tsunami is overdue (LANDER & WHITESIDE, 1997). The risk of potentially destructive tsunamis is greatest from the West Indies island arc with its sub-aerial and submarine volcanoes, steep underwater slopes and the active subduction zones of the Caribbean Plate boundary marked by numerous earthquakes. We are still some way from realistic assessments of risk at vulnerable Caribbean coastlines with their dense coastal population centers, harbor infrastructures, petroleum carriers and hotel and beach facilities, nevertheless we can suspect that tsunamis of similar size like the observed paleotsunamis would devastate the low-lying coastal infrastructures to a large amount. In the near future further efforts should concentrate on geomorphologic field studies on a Caribbean-wide scale to understand the nature of tsunami deposits and to precise and extend the existing Caribbean Tsunami catalogue comprehended so far by LANDER & WHITESIDE (1997).

With regard to the results of this study it must be stressed with great emphasis that the establishment of a feasible and effective Intra-Americas Sea Tsunami Warning System as it is visualized by the Intergovernmental Oceanographic Commission of the UNESCO is an important step to mitigate future disasters. We hope to encourage with this study the governmental institutions on a local and a Caribbean-wide scale to intensify activities in tsunami related education, warning, management as well as research, as for the Caribbean the rare opportunity exists to establish a warning system before another disastrous tsunami occurs, unlike in other regions where most warning systems have been established following a disaster.

List of Figures

Fig. 1: The islands of Aruba, Curaçao and Bonaire in the Southern Caribbean.

Fig. 2: Relief and localities of Aruba.

Fig. 3: Relief and localities of Bonaire.

Fig. 4: Relief and localities of Curaçao.

Fig. 5: Temperatures and precipitation on Curaçao (Hato Airport, 8 m asl).

Fig. 6: Nearly all exposed vegetation on the islands of Aruba, Curaçao and Bonaire (except of the cacti) show a significant adaptation in their growth form to the strong and persistent easterly trade winds. Divi-divi tree at Boka Washikemba, Bonaire.

Fig. 7: Main structural features and geodynamic situation in the Caribbean: Plate boundaries, transform faults, subduction zones and volcanic island arcs (after SCHUBERT, 1988; MANN et al., 1990).

Fig. 8: Bathymetric map of the Caribbean Sea: A deep-sea basin of -4000 m to -5000 m forms most of the area. The eastern border is formed by the Lesser Antilles island arc system, which is the expression of the subduction of the Atlantic plate beneath the Caribbean plate. Most of the active volcanoes are located in the central part of the arc. The islands of Curaçao and Bonaire rise from a depth of more than 2000 m. They are separated from the South American mainland by the Bonaire basin, while Aruba is situated on the Venezuelan shelf.

Fig. 9: Geology of Aruba: Pleistocene fringing reefs of at least three different isotope stages surrounding a non-carbonate volcanic basement (after DE BUISONJÉ, 1974).

Fig. 10: Geology of Bonaire: Pleistocene fringing reefs of at least three different isotope stages surrounding a non-carbonate volcanic basement (after DE BUISONJÉ, 1974).

Fig. 11: Geology of Curaçao: Pleistocene fringing reefs of at least three different isotope stages surrounding a non-carbonate volcanic basement. (after DE BUISONJÉ, 1974).

Fig. 12: Schematized geological evolution of Curaçao.

Fig. 13: Geological cross-section of the modern island of Curaçao. The cross-section illustrates the elevation of depositional and erosional terraces of the Seroe Domi Formation and Quaternary Limestone Terraces (= fossil coral reefs) (DE JONG, 2000).

Fig. 14: Paleo-cliff undercut at about 10 m asl by a deep bioerosive notch formed during the sealevel highstand of the last interglacial (isotope stage 5e) in the Middle Terrace near Hato Airport, Curaçao.

Fig. 15: Sealevel rising trend for the Caribbean coasts of Venezuela in calendar years before present (RULL, 2000).

Fig. 16: One of the typical ria-like embayments at the leeward coast of Curaçao (Spaanse Water).

Fig. 17: The small and narrow embayments with steep side walls are called "boka" on the Antillean islands. They mostly appear at the windward coasts (Boka Djegu in Shete Boka Park, Curaçao).

Fig. 18: Four different types of bioerosive notches on the islands of Aruba, Curaçao and Bonaire are strictly depending on the surf environment (after FOCKE, 1978c).

Fig. 19: The distribution of wave heights, cliff types and notches along the coastlines of Curaçao and Bonaire: A distinct difference can be seen between the windward (NE) and the leeward coasts (SW).

Fig. 20: Example of a deep incised bioerosive notch at the leeward to lateral coastal section of Curaçao. No signs of sealevel variations during notch formation can be identified.

Fig. 21: Between Sero Colorado and Boka Grandi at the exposed windward coastline of Aruba a very well developed bench can be seen with a width of more than 10 m. It clearly points to a long lasting sealevel during the Holocene.

Fig. 22: Typical micro-relief in the supra-tidal spray zone on carbonate rocks along the exposed coastlines of Aruba, Curaçao and Bonaire: up to one meter deep rock pools carved by the grazing activity of gastropods (mostly littorinids), separated by very sharp edges. No sediments are present in these environments.

Fig. 23: Front, surface and upper parts of the benches in the very exposed environments of the intertidal zone on Aruba, Curaçao and Bonaire often show a pattern of vermetid rims separating shallow pools. The vermetid accretions can be developed up to 2 m above mean high water level and are very resistant against mechanical destruction by surf beat. They protect the underlying reef rock from bioerosion.

Fig. 24: The development of a bench along a steep and sediment free exposed coastline in carbonate rocks of the Antillean islands. Accretions of calcareous algae and vermetids protect a certain part of the rock in the intertidal zone against bioerosion, whereas above and below this protected zone the cliff front is cut back by bioerosion.

Fig. 25: A typical landscape aspect of the islands. Columnar cactus (*Cereus repandus*) tower over thorn scrub vegetation.

Fig. 26: The largest stand of mangroves (*Avicennia* sp. and *Rhizophora* sp.) on the islands can be found in Lac Baai in the southeast part of Bonaire.

Fig. 27: Saltpans cover extended areas of former lagoons and low lying Pleistocene reef flats in Bonaire.

Fig. 28: Mining activities of coral and limestone debris (tsunami deposits) in Curaçao and Bonaire. **A:** Spelonk area, Bonaire (aerial photograph). **B:** Hato area, Curaçao (aerial photographs).

Fig. 29: Oil terminal in Aruba. The refineries of the islands have been constructed in close vicinity to the coast, all with their own terminals for large tankers.

Fig. 30: Coastal infrastructure on Bonaire. The infrastructure along extended stretches of the leeward coastlines on all islands (hotels, apartments, marinas, roads, etc.) show very high investments in a zone, which has been reached by strong tsunamis during the Holocene.

Fig. 31: Tsunami velocity in relation to water depth. Tsunamis travel at several hundreds kilometers per hour at a speed proportional to the square root of the water depth. Consequently, they can travel across an ocean basin the size of the Pacific in less than 24 hours.

Fig. 32: The breaking point of waves is depending on wave height and water depth. Along the north coast of Curaçao all waves >12 m are breaking at the front of the subtidal platform in a distance of 120 m. At the cliff front all waves >4 m will break. Tsunami waves, however, can surge across dry beds or shores.

Fig. 33: The wave of the 1960 Chile tsunami was observed all around the island of Hawaii. Note the differences and variations in wave heights over small distances (after: COX & MINK, 1963).

Fig. 34: Reported runup heights for the Alaska Tsunami from 1946 on Oahu (after: COX & MINK, 1963).

Fig. 35: Tsunami impacts since 1900 AD. The coastlines of the world, which are mostly affected by recent tsunamis are concentrated around the Pacific basin. Here, tsunamis are mainly caused by subduction earthquakes. Another area with frequent tsunami occurrence is located in the Eastern Mediterranean.

Fig. 36: Tsunami impacts covering the period from 1500-1899 AD. Extending the observed tsunami records in the past two trends become obvious: The frequency of tsunami impacts increases in most regions from single events to multiple events and areas hitherto unknown for their potential tsunami risk can be found on the map.

Fig. 37: Tsunami impacts covering the period from 3000-500 BP. Further back in time, the geographical distribution of observed tsunami impacts gets wider, the more sedimentary records can be deciphered.

Fig. 38: The wind and pressure profiles through hurricane *Anita* clearly show the distinct areas of the hurricane eye associated with low pressure and the ring with the highest wind velocities tangent to the eye wall (Source: FLETCHER *et al.*, 1995).

Fig. 39: The profiles through hurricanes with different intensities illustrate that the band of highest wind velocities becomes more narrow and distinct with higher hurricane intensity (Source: SIMPSON & RIEHL, 1981).

Fig. 40: The steering winds and the winds associated with the anti-clockwise motion add together on the right side of the tropical circulation system. In the upper right quadrant the strongest winds and highest waves are generated (Source: SCHEFFERS & KELLETAT, 2001).

Fig. 41: Hurricane *Celia* reaches wind velocities up to 250 km/h. The bandwidth of the gusts fluctuates between 40 - 100 km/h. The wind velocity drops from 256 km/h in the eye wall to 15 km/h in the center of the eye (after: SIMPSON & RIEHL, 1981).

Fig. 42: The highest storm surges are located on the right side of a tropical circulation system of the Northern Hemisphere and coincide with the highest wind velocities (after: SIMPSON & RIEHL, 1981).

Fig. 43: To generate a wave height of 12.2 m a wind velocity of 120 km/h (category 1 hurricane) over a period of 15 hours with a fetch of approximately 148 km is necessary. The lower the wind velocities, the longer the duration of the hurricane associated wind fields have to last to produce equivalent wave height (after: SIMPSON & RIEHL, 1981).

Fig. 44: Relation between the wave height and the energy of a wave in 1 m steps. The increase in wave heights with regard to the geomorphic effects plays a more important role in lower wave heights categories, whereas an increase of wave height in higher wave classes is of less geomorphic relevance.

Fig. 45: Relation between wave heights and the distance to the area of wave origin and the origination of oceanic swell (after: SIMPSON & RIEHL, 1981).

Fig. 46: Different approach of tsunami and storm waves at a cliff coast.

Fig. 47: The factors, which influence the movement of boulders in nature are extremely various. **A:** 29 parameters illustrate exemplary the complexity of the boulder movement process. **B:** The bandwidth of possible variations for some factors.

Fig. 48: Relation between the weight of a boulder and its shape – round or cubic. The weight of a cubic boulder increases rapidly with increased diameter.

Fig. 49: Illustration of boulder movement and measuring

List of Figures

Fig. 50: method to calculate boulder volume and weight. Exemplary field evidences of boulder deposition by hurricanes or tsunamis.

Fig. 51: Tracks of tropical storms over the North Atlantic Area from 1886 to 1995. In total 956 storms were recorded (METEOROLOGICAL SERVICE OF THE NETHERLANDS ANTILLES AND ARUBA, 1998). The density of the tracks illustrates the high frequency of hurricanes in this geographical region. During major hurricanes (categories 3, 4 or 5 on the SAFFIR-SIMPSON Hurricane Scale) like *Gilbert* (1988), *Hugo* (1989) and *Luis* (1995) wind speeds exceeding 300 km/h were observed.

Fig. 52: The landfall of hurricanes during the time period 1935-1965. The coastlines along the northern Caribbean, the Golf of Mexico and Florida are more affected by hurricane landfalls than the southern Caribbean, where landfalls are very rare (after LANDSEA, 1993).

Fig. 53: Historical tsunamigenic events (46) in the Caribbean, not all events are present because 11 events lack their source coordinates (Source: LANDER & WHITESIDE, 1997). The size of the symbol is proportional to the event magnitude.

Fig. 54: Only few hurricanes passed within 100 nm from Curaçao, Bonaire and Aruba over the time period from 1605 to 1998

Fig. 55: Hurricane *Lenny*'s pass through the Caribbean. The system was formed south of Jamaica and moved eastwards towards the Lesser Antilles. The ABC- islands experienced heavy surf due to stronger wind velocities and wave height on the right side of its track (after: GUINEY, 2000).

Fig. 56: Main tsunami depositional types and their genetic source in the coastal environment.

Fig. 57: Cross-section of a typical hurricane and tsunami ridge.

Fig. 58: Debris ridge (nearly 1 m high) consisting chiefly of rods of *Acropora cervicornis* branches with tongues of shingle. Pink Beach, leeward coast, Bonaire.

Fig. 59: Subrecent debris ridge at Willemstoren, leeward coast of Bonaire. The steep seaward margin locally reaches 80° and the gentle sloping inward face may show angles of less then 10°.

Fig. 60: Subrecent debris ridge showing a steep seaward flank and a more gentle dipping leeward margin. The light colors on the seaward basis of the ridge show the reworking and leeward migration of the ridge mainly by more frequent lower magnitude events.

Fig. 61: Rampart formation at Dos Boka, windward coast of Curaçao, located at +6 m asl and about 40 m distant from the cliff front.

Fig. 62: Boulder arrangement at the windward coast of Bonaire, Spelonk Lighthouse. One of the last areas, which is not disturbed by mining activities and show the original distribution of limestone boulders deposited by a high magnitude event.

Fig. 63: Localization of all figures for Aruba.

Fig. 64: Small islands at the leeward side of Aruba showing older weathered ridges and some recent, light coral rubble from hurricane *Lenny* (1999).

Fig. 65: Shallow Holocene reef flat with extensive sand sheets in front of the beaches at the west coast of northern Aruba, mostly inhabited by seagrass fields. Living coral reefs are not abundant.

Fig. 66: The broad spit of Manchebo beach at the west coast of Aruba with wide beaches built up by storm waves from coral sand and shingle of the extensive submarine platform.

Fig. 67: South west of the California sand dunes in the north of Aruba fresh smaller coral fragments and weathered older and bigger ones are deposited on top of low cliffs.

Fig. 68: The ridge of coral debris deposited by hurricane *Lenny* (Nov. 1999) covering partly coarser and older coral boulders.

Fig. 69: At the northernmost tip of Aruba an old weathered rampart of coral debris has partly been covered by vegetation. In distance, the California sand dunes can be seen.

Fig. 70: Oblique aerial photograph of a strip of boulders and partly destructed benches at the northern tip of Aruba island.

Fig. 71: Oblique aerial photograph of northern Aruba (view to southeast). The strictly east-west-shifting California sand dunes partly cover a broad and old boulder rampart seen as a belt of dark gray colors.

Fig. 72: East of the California sand dunes, a scattered field of boulders is deposited up to 200 m from the cliff front.

Fig. 73: Huge boulders from a low cliff, up to 10 t in weight. Sand deposits in the shelter of the boulders generate the light colors in the area. In the distance the California lighthouse.

Fig. 74: Landward slope of the weathered boulder ridge near Druif, windward coast of northern Aruba. Grass vegetation is settling on the lower parts of the landward flank of the ridge.

Fig. 75: The old boulder ridge south of Druif contains rounded material as well as well preserved corals (*Diploria* sp.) and fragments of vermetid benches (with hammer).

Fig. 76: At the seaward slope of the older boulder ridge bigger fragments as well as sand deposits originating from the lowermost strata are situated. South of Druif.

Fig. 77: Lines of boulders or ramparts follow the contours of the cliff line, but have been slightly shifted aside due to wave impact approaching from left (the east). Central windward coast of Aruba. Oblique aerial photograph.

Fig. 78: Many of the rampart formations along the windward side of Aruba have been removed by mining. Oblique aerial photograph.

Fig. 79: Oblique aerial photograph of the east part of Andicuri Bay, south of Natural Bridge, central east coast of Aruba. Boulders on top of a Pleistocene coral reef flat lie at +9 m asl The reddish colors show remnants of a fossil soil covered with boulders and rampart debris.

Fig. 80: Top of the Andicuri rampart and ridge with boulders up to 6 tons at 9 m asl The black rounded fragments are diabas boulders derived from the foreshore environment.

Fig. 81: Detailed look at the ridge deposits of Andicuri Bay at 9 m asl: well-rounded coral and diabas pebbles and cobbles, a coral head of *Diploria strigosa*, and many broken *Cittarium pica* shells.

Fig. 82: Aerial view of Boka Prins: windblown sand sheets and remnants of a boulder rampart between the two bays. See the asymmetrical shape of the bays due to wave impact from left (east).

Fig. 83: Boka Grandi with beaches and vegetated dunes in front of a fossil cliff in a Pleistocene coral reef terrace. The surf aspect shows shallow water conditions behind a fringing Holocene coral reef.

Fig. 84: Sero Colorado (left), the most southeastern point of Aruba island with the attached lower coral reef terrace to the east. Note the perfectly developed bench. The gray colors of the old reef flat show scattered rampart deposits.

Fig. 85: Close-up oblique aerial view of the bench and ramparts north of Sero Colorado.

Fig. 86: Terrestrial view along the bench north of Sero Colorado. Scattered smaller boulders on the left in distance to the cliff at around +5 to +7 m asl.

Fig. 87: Double ridges west of Sero Colorado in front of the lowest Pleistocene coral reef terrace.

Fig. 88: Coral sand and weathered *Acropora* boulders building the double ridge west of Sero Colorado. The vegetation shows that the deposition has been taken place at least many decades ago.

Fig. 89: Distribution of coarse Holocene littoral deposits on Aruba island left by young hurricane *Lenny* or by older tsunamis.

Fig. 90: Three cross-sections (NW, SE and Central Part) over the island of Curaçao.

Fig. 91: Location of figures on Curaçao.

Fig. 92: Distribution of coarse Holocene littoral deposits on Curaçao left by recent hurricane *Lenny* or older subrecent tsunamis.

Fig. 93: Aerial view of Awa di Oostpunt with the spit-like ridge enclosing the inner bay area.

Fig. 94: Weathering front penetrates the debris deposits at Fuikbaai only for some centimeters.

Fig. 95: Rubble ridge to the west of Awa di Oostpunt showing a steep crest and a convex profile.

Fig. 96: Oblique aerial picture of the double ridges between Awa di Oostpunt and Awa Blanku. The arrows mark the two distinguished deposits, which are separated by a narrow line of vegetation.

Fig. 97: Oblique aerial view of Spaanse water, a ria-like inland bay, which is intensively developed for housing projects and a private harbor.

Fig. 98: Oblique aerial photograph of eastern Caracasbaai, with the subrecent ridge deposits on top of the Lower Terrace at a height of +3.5 m asl.

Fig. 99: The subrecent ridge deposits at Lagun Jan Thiel consisting chiefly of rods of *Acropora* branches.

Fig. 100: The distribution of shape classes showing the source of the debris material in percent. Over 90% derive from the Holocene reef environment.

Fig. 101: Small notch opposite the Sea Aquarium indicating a Holocene sealevel dislocation of 0.4 m. This location shows the only evidence of a Holocene dislocation we observed on the Leeward Lesser Antilles.

Fig. 102: *Lenny* deposits on the coastal section behind the water factory. The barrier ridge is colonized by mangrove species.

Fig. 103: Largest boulder perched on the leeward coast of Curaçao at the domain of the Sheraton Hotel. This boulder is 6.5 × 4.6 × 2.6 m in size with a weight of 194 tons.

Fig. 104: Boulder assemblage around the remnants of a small ruin belonging to Fort Piscadera, which was built in the 17th century.

Fig. 105: Boulders resting on the Lower Terrace deposits at a height of +3.0 m asl at Slangenbaai.

Fig. 106: At the western side of Bullenbaai extensive rubble ridges have been deposited and are now colonized by shrub vegetation indicating a subrecent age of the rubble.

Fig. 107: Debris accumulations in front of a paleocliff at Bullenbaai. The deposits show imbrication and have an upward sloping profile in direction of the paleocliff.

Fig. 108: Diagram of shape measurements at Bullenbaai.

Fig. 109: Ridge of fresh coral rubble (mostly *Acropora cervicornis*) from Hurricane *Lenny* (Nov. 1999) with a maximum height of 0.6 m above high wa-

List of Figures

Fig. 110: ter level, partly affecting living vegetation. The picture was taken 90 days after the event.

Fig. 110: Caracasbaai with the Seroe Mansinga at the left side of the picture. Note the almost vertical escarpment marking the rupture of the rock mass, which slided down during the Caracasbaai event. In front, a sequence of shingle beaches has been built up by storm activities.

Fig. 111: Bathymetric map of Caracasbaai. The isobaths trace the submarine scar, which is excavated from the sea bottom. The scar is situated about 150 m below the surrounding level of the sea bottom and has steep sides (after: DE BUISONJÉ & ZONNEFELD, 1976).

Fig. 112: Fathometer profile, south of Caracasbaai, taken during the submersible dive of the Harbor Branch Oceanographic Institution on May 9, 2000. At the depth between 700-800 m two pinnacles, each up to 60 m high and 550 m in diameter, which slumped form Caracasbaai, have been surveyed (Source: REED, 2001).

Fig. 113: Overview of the geomorphologic situation of Caracasbaai, south coast of Curaçao. Huge boulders (the smallest with several 1000 metric tons) are located in the innermost part of Caracasbaai. Several clusters of boulders showing their former attachment to a single block mass. Fort Beekenburg, a 17th century fortification is build on one of the blocks.

Fig. 114: Oblique aerial view of the Watamula-Noordpunt area with the transition between the more sheltered lateral (right) and the windward cliff profile (left). Some huge boulders, a rampart and a ridge-like feature can be seen on top of the Lower Reef Terrace.

Fig. 115: Huge tsunami boulders at Watamula, Curaçao.

Fig. 116: The largest Watamula boulder has been overturned and the rock pools of the supratidal are now situated at its base.

Fig. 117: Reworked ridge as the front part of an older rampart at Noordpunt, Curaçao. In front of the ridge, a band of shelly sand is sedimented, obviously washed out from the basis of the ridge.

Fig. 118: Perfectly preserved rampart near Dos Boka with very coarse debris approximately 70 m distant from the cliff at +8.0 m asl.

Fig. 119: Illustration of ripple marks formed in coarse rampart deposits by very strong currents at more than 8.0 m asl, west of Dos Boka.

Fig. 120: Aerial view of the rampart formation at Dos Boka. Note the ripple marks (see arrow).

Fig. 121: The rampart deposition on both sides of nearly all bokas in the northern part of Curaçao show an asymmetric pattern with the nearly absence of debris along the western flanks.

Fig. 122: Bare karst surface of the Lower Terrace in front of the rampart near Un Boka.

Fig. 123: Typical rock pools up to 1.0 m deep lacking coarse sediments in the supratidal belt on top of the cliff on the Lower Reef Terrace.

Fig. 124: A thin organic platform constructed by vermetids and calcareous algae is characteristic for the entrance areas of bokas.

Fig. 125: Diagram showing the percentage of coarse particles in the rampart near Un Boka in 7 different shape classes.

Fig. 126: Orientation of the longest axis of debris fragments on 6 sites in the rampart near Dos Boka, Curaçao.

Fig. 127: Even large boulders may still show perfectly preserved rock pools and bioerosive sculpturing despite transportation hundred meters inland.

Fig. 128: Very coarse rampart debris in a scattered layer close to Oostpunt, Curaçao.

Fig. 129: At Oostpunt the largest tsunami boulder of all three islands can be seen. The weight amounts to 281 t.

Fig. 130: Map of the distribution of ramparts and boulders near Oostpunt, Curaçao.

Fig. 131: The narrow rampart along a part of the leeward coast of Klein-Curaçao.

Fig. 132: Overview of the figures described on Bonaire.

Fig. 133: Distribution of tsunami and hurricane *Lenny* debris along the coasts of Bonaire and Klein Bonaire.

Fig. 134: Fresh looking, wide ridge of coral rubble north of Pink Beach, formed during *Lenny* in 1999.

Fig. 135: Orientation of debris on the hurricane *Lenny* ridge at Pink Beach, Bonaire.

Fig. 136: Boulders of up to 2 tons slightly moved onshore by hurricane *Lenny* at Punt Vierkant, Bonaire.

Fig. 137: Old storm ridge deposit east of Landhuis Karpata, Bonaire.

Fig. 138: The barrier at the entrance of Salina Tern, formed during a subrecent tsunami event.

Fig. 139: Evidences of hurricane *Lenny* along a coastal stretch of about 1 km north of Slagbaai. The illustration documents the pre-*Lenny* conditions with reference to an aerial photograph of 1996 and the post-*Lenny* situation by oblique aerial photos and terrestrial pictures (Jan. 2001) as well as mapping results of the onshore deposits.

Fig. 140: A small embayment north of Boka Slagbaai is filled with *Lenny* deposits up to a height of 5.0 m asl.

Fig. 141: Destroyed houses at the northern end of Boka Slagbaai, Bonaire.

Fig. 142: High waves hit a plantation house of 1876, destroying the seaward walls until the height of the ridge. Part of the walls were transported further inland.

Fig. 143: At smaller houses at the southern end of Slagbaai, the direction of the wave impact can be seen: All walls at the seafront were destroyed, and even the backside walls have been collapsed. Only walls perpendicular to the coast have offered resistance and are still in place.

Fig. 144: Aerial view of the recently formed coral rubble spit by hurricane *Lenny*.

Fig. 145: Oblique terrestrial photograph of the spit attached to the pre-existing shoreline.

Fig. 146: On top of a 5.0 m high cliff a perfect semicircle of fresh coral debris has been accumulated by hurricane *Lenny*.

Fig. 147: Oblique aerial view of the semicircle debris deposits controlled by wave refraction during hurricane *Lenny*.

Fig. 148: Orientation of coarse debris on the tsunami ridge at Boka Bartol.

Fig. 149: Shape of the tsunami ridge deposits at Boka Bartol and those of the seaward reworked part of the ridge by hurricane *Lenny* in 1999.

Fig. 150: The southern shores of Klein Bonaire with an old tsunami ridge.

Fig. 151: The western spit of Klein Bonaire: from the north, the recent storm ridge of hurricane *Lenny* has been attached outside to an older tsunami ridge deposit.

Fig. 152: Boka Kokolishi with perfectly semicircle grown vermetid and algal rims. Note the boulder of nearly 100 tons on top of the Lower Reef Terrace.

Fig. 153: Oblique aerial photograph of large boulder assemblages in front of the Middle Terrace of Seru Grandi in Washington Nationalpark, Bonaire.

Fig. 154: Boulder of more than 100 t located in Washington Nationalpark in the area of Seru Grandi.

Fig. 155: Boulder of more than 100 tons deposited 100 m distant to the cliff front.

Fig. 156: Assemblage of two boulders with a weight of about 100 t.

Fig. 157: Partly cemented tsunami rubble on top of a cliff south of Playa Chikitu. In the distance, a huge tsunami boulder 9.20 m long with a weight of 120 t is visible.

Fig. 158: Boulder assemblage at Boka Olivia, one presumably broken apart during transportation and deposition.

Fig. 159: An impressive boulder field south of Spelonk Lighthouse, situated in rather dense vegetation more than 150 m apart from the shoreline.

Fig. 160: Four examples of huge boulders deposited in the area south of Spelonk, each with a weight of over 100 tons.

Fig. 161: Detailed map of the northern reference area at Spelonk Lighthouse.

Fig. 162: Detailed map of the southern reference area 3 km south of Spelonk Lighthouse.

Fig. 163: These two boulders of 126 and 137 t have been moved by the tsunami in one piece and settled down about 160 m distant to the cliff at a height of +5.0 m asl. During the deposition, the boulder was broken and the fragments separated for several meters. The breaking of the two pieces can clearly be demonstrated by the fitting of large coral colonies in both of them.

Fig. 164: Very well preserved rock pools at a boulder 120 m distant to the shoreline.

Fig. 165: Extensive mining has removed most of the debris and exposed the unweathered surface of the Lower Terrace, which show initial soil formation before the tsunami events documented with the brownish colors.

Fig. 166: Heavy equipment in modern times leads to complete devastation of the landscape and a more accurate pattern of the residual deposits.

Fig. 167: Sharp contrast between old undisturbed rampart and fresh mining areas south of Lagoen.

Fig. 168: The ramparts south of Lagoen clearly show two parallel units with slightly different color and therefore presumably different age.

Fig. 169: The broad and coarse tsunami ridge deposits along the southeastern coastline of Bonaire.

Fig. 170: Seaside vertical section of the tsunami ridge in southern Bonaire showing a weathered coarse upper layer with fresh looking sand at the basis.

Fig. 171: Diagram of situmetric measurements, southeastern Bonaire.

Fig. 172: Diagram of shape measurements, southeastern Bonaire.

Fig. 173: Weight, transport distance and depositional height of boulder deposits for the ABC-islands.

Fig. 174: Average weight of the largest boulder on Aruba, Curaçao and Bonaire.

Fig. 175: Diagram of boulder weight and depositional distance for boulder deposits of the ABC-islands.

Fig. 176: Diagram representing the energy level of the impact derived from boulder weight, transport distance and depositional height.

Fig. 177: A tsunami approaching Curaçao. The arrival times reflecting the real values for water depth around the island.

Fig. 178: Overview of impressive paleotsunami imprints on the ABC-islands.

Fig. 179: Age distribution of all 43 dated samples. The conventional radiocarbon ages are visualized as single line, the calibrated (2s) ages are presented in form of triangles. Reservoir age = 429 years.

Fig. 180: Absolute age distribution related to the different species.

Fig. 181: Distribution of absolute ages in relation to the different debris formations.

Fig. 182: Comparison of the absolute radiocarbon ages of all three islands.

Fig. 183: Location of the ^{14}C-samples and conventional radiocarbon ages on Aruba.

Fig. 184: Location of the ^{14}C-samples and conventional radiocarbon ages on Bonaire.

Fig. 185: Location of the ^{14}C-samples and conventional radiocarbon ages on Curaçao.

Fig. 186: Suggested direction of paleotsunamis impacting the ABC-islands.

List of Tables

Tab. 1: Geographical data on Aruba, Curaçao and Bonaire (DE JONG, 1999).

Tab. 2: Climate data of Curaçao for the years 1971-2000, Hato Airport.

Tab. 3: The IMAMURA-IIDA Tsunami Magnitude Scale.

Tab. 4: The tsunami intensity scale after SOLOVIEV (1978).

Tab. 5: The Saffir-Simpson Hurricane Scale.

Tab. 6: The main characteristics of hurricane and tsunami waves as reported by field observations.

Tab. 7: Relation between wave height and the weight of boulders moved by this particular wave height (LORANG, 2000).

Tab. 8: Relation between wave height, wave period and the movement of boulders (OAK, 1985).

Tab. 9: Dimensions of selected boulders on the ABC-islands and height of waves (tropical cyclone and tsunami) required to overturn these boulders. For the calculation of wave heights the equations of NOTT (1997) were applied.

Tab. 10: Historical Caribbean Tsunami Record from 1530 to 1991 (LANDER & WHITESIDE, 1997).

Tab. 11: Characteristics of hurricane and tsunami debris deposits.

Tab. 12: List of large tsunami boulders from different sites along the shoreline of Aruba.

Tab. 13: List of large tsunami boulders from different sites along the shoreline of Curaçao.

Tab. 14: List of large tsunami boulders from different sites along the shoreline of Bonaire.

Tab. 15: Conventional radiocarbon ages for 45 samples from Aruba, Curaçao and Bonaire.

Tab. 16: Values for the reservoir age and Delta R for the Caribbean, Florida, Bahamas, Tortugas and Bermuda (Radiocarbon Lab., Queen's University of Belfast, this study).

Tab. 17: Calibrated absolute ages for 43 samples from Aruba, Curaçao and Bonaire, sorted after the conv. ^{14}C age (calibrated with MARINE98 and CALIB4 after STUIVER, REIMER & BRAZIUNAS, 1998), Reservoir age = 429 years.

Tab. 18: Absolute ages and required minimum depositional wave heights for four measured boulders (for formulas see Chapter 3).

References

ABE, K. (1973): Tsunami and mechanism of great earthquakes.- Physics of Earth and Planetary Interiors, **7**: 143-153.

ABE, K. (1979): Size of great earthquakes of 1873-1974 inferred from tsunami data.- Journal of Geophysical Research, **84**: 1561-1568.

ABE, K. (1981): Physical size of tsunamigenic earthquakes of the northwestern Pacific.- Physics of Earth and Planetary Interiors, **27**: 194-205.

ABE, K. (1989): Quantification of tsunamigenic earthquakes by the M_t scale.- Tectonophysics, **166**: 27-34.

ALEXANDER, C.S. (1961): The marine terraces of Aruba, Bonaire and Curaçao, Netherlands Antilles.- Annals of the Association of American Geographers, **51**(1): 102-123.

ANTHES, R.A. (1982): Tropical Cyclones, Their Evolution, Structure and Effects.- American Meteorological Society, 208 pp., Boston.

AUDEMARD, F.A. (2001): Quaternary tectonics and present stress tensor of the inverted northern Falcón Basin, northwestern Venezuela.- Journal of Structural Geology, **23**: 431-453.

BAINES, G.B.K., BEVERIDGE, P.J. & MARAGOS, J.E. (1974): Storms and island building at Funafuti atoll, Ellice Islands.- Proceedings Second International Coral Reef Symposium, **2**: 485-496.

BAINES, G.B.K. & MCLEAN, R.F. (1976): Sequential studies of hurricane deposit evolution at Funafuti Atoll.- Marine Geology, **21**: M1-M8.

BAK, R.P.M. (1975): Ecological aspects of the distribution of coral reefs in the Netherlands Antilles.- Bijdragen tot de Dierkunde, **45**(2): 181-190.

BAK, R.P.M. (1977): Coral reefs and their zonation in the Netherlands Antilles.- In: Frost, S.H., WEISS, M.P. & SAUNDERS, J.B. (eds.): Reefs and related Carbonates: Ecology and Sedimentology.- American Association of Petroleum Geologists, Studies in Geology, **4**: 3-16, Tulsa.

BALL, M., SHINN, E. & STOCKMAN, K. (1967): The geological effects of hurricane Donna in South Florida.- Journal of Geology, **75**: 583-597.

BANDOIAN, C.A. & MURRAY, R.C. (1974): Pliocene-Pleistocene carbonate rocks of Bonaire, Netherlands Antilles.- Geological Society of America Bulletin, **85**: 1243-1252.

BASCOM, W. (1959): Ocean Waves.- In: MOORE, J.G. (ed.): Oceanography: 45-55; Reprints from Scientific American 1971, San Francisco.

BEERS, C.E., DE FREITAS, J. & KETNER, P. (1997): Landscape Ecological Vegetation Map of the Island of Curaçao, Netherlands Antilles.- Publications Foundation for Scientific Research in the Caribbean Region: **138**; 51 pp.

BEETS, D.J. (1972): Lithology and Stratigraphy of the Cretaceous and Danian Succession of Curaçao.- Publications Foundation for Scientific Research in the Caribbean Region, **70**; 153 pp.

BEETS, D.J. & MAC GILLAVRY, H.J. (1977): Outline of the Cretaceous and Early Tertiary History of Curaçao, Bonaire and Aruba.- Field Guide 8th Caribbean Geological Conference (Curaçao), **15**: 1-6.

BEETS, D.J., MARESCH, W.V., KLAVER, G.T., MONTTANA, A., BOCCHIO, R., BEUNK, F.F. & MONEN, H.P. (1984): Magmatic rock series and high pressure metamorphism as constraints on the tectonic history of the Southern Caribbean.- In: BONINI, W.B. HARGRAVES, R.B. & SHAGAM, R. (eds.): The Caribbean-South America Plate Boundary and Regional Tectonics.- Geological Society of America Memoirs, **162**: 95-130.

BEVINGTON, P.R. (1969): Data reduction and error analysis in the Physical Sciences.- McGraw-Hill Book Company.

BLUMENSTOCK, D.I., FOSBERG, F.R. & JOHNSON, C.G. (1961): The re-survey of typhoon effects on Jaluit atoll in the Marshall Islands.- Nature, **189**: 618-620.

BOERSTRA, E.H.J. (1982): De precolumbiaanse bewoners van Aruba, Curaçao and Bonaire.- De Walberg Pers, Zutphen: 79 pp.

BOLDINGH, I. (1914): The flora of the Dutch West Indian Islands. II. The flora of Curaçao, Aruba and Bonaire: 197 pp.; E.J.Brill, Leiden, The Netherlands.

BOURROUILH-LE JAN, F.G. & TALANDIER, J. (1985): Sédimentation et fracturation de haute énergie en milieu récifal: tsunamis, ouragans et cyclones et leurs effets sur la sédimentologie et la géomorphologie d'un atoll: Motu et Hoa, à Rangiroa, Tuamotu, Pacifique SE.- Marine Geology, **67**: 263-333.

BRANDSMA, M., DIVOKY, D. & HWANG, L.-S. (1974): Response of small islands to long waves.- Tetra Tech. Inc., California, for U.S. Atomic Energy Commission, Nevada.

BROECKER, W.S. & OLSON, E.A. (1961): Lamont radiocarbon measurements VIII.- Radiocarbon, **3**: 176-204.

BRUUN, P. (1990): Port Engineering.- Gulf Publishing, Houston.

BRUUN, P. (1994): Freak Waves in the Ocean and Along Shores, Including Impacts on Fixed and Floating Structures.- In: FINKL, CH.F. (ed.): Coastal Hazards. Perception, Susceptibility and Mitigation.- Journal of Coastal Research, Special Issue **12**: 163-175; Fort Lauderdale.

BRYANT, E.A., YOUNG, R.W. & PRICE, D.M. (1992): Evidence of tsunami sedimentation on the southeastern coast of Australia.- Journal of Geology, **100**: 753-765.

BRYANT, E.A., YOUNG, R.W. & PRICE, D.M. (1996): Tsunami as a Major Control on Coastal Evolution, Southwestern Australia.- Journal of Coastal Research, **12** (4): 831-840.

BUGGE, T., BELDERSON, R.H. & KENYON, N.H. (1988): The Storegga Slide.- Philosophical Transactions of the Royal Society of London, Ser. A, **325**: 357-388.

CAMFIELD, F.E. (1993): Dynamic response of structures to tsunami attack.- In: TINTI, S. (ed.): Tsunamis of the World: 133-138; Kluwer, Dordrecht.

CAMFIELD, F.E. (1994): Tsunami Effects on Coastal Structures.- In: FINKL, Ch.F. (ed.): Coastal Hazards. Perzeption, Susceptibility and Mitigation.- Journal of Coastal Research, Special Issue **12**: 177-187; Fort Lauderdale.

CAPUTO, M. & FAITA, G. (1984): Primo Catalogo Dei Maremoti Della Coste Italiane.- Atti Accademia Nazionale dei Lincei, Memorie, Classe Science Fisiche Matematiche Naturali, Vol. **XVII**: 213-356.

CARRACEDO, J.C., DAY, S.J., GUILLOU, H. & TORRADO, F.J.P. (1999): Giant Quaternary landslides in the evolution of La Palma and El Hierro, Canary Islands.- Journal of Volcanology and Geothermal Research, **94**: 169-190.

CBS (Centraal Bureau voor de Statistiek, N.A.) (1996): Economisch profiel, Nederlandse Antillen.- CBS, Curaçao.

CHAGUÉ-GOFF, C. & GOFF, J. (1999): Paleotsunami: Now you see them, now you don't.- Tephra, **10**: 10-12.

COX, D. & MINK, J.F. (1963): The Tsunami of 23 May 1960 in the Hawaiian Islands.- Bulletin of the Seismological Society of America, **53**(6): 1191-1209.

COX, D.C. & PARARAS-CARAYANNIS, G. (1976): Catalog of tsunamis in Alaska revised 1976.- Report SE-1, 43 pp., World Data Center A, NOAA; Boulder.

DAVIES, P. & HASLETT, S.K. (2000): Identifying storm and tsunami events in coastal basin sediments.- Area, **32**(3): 335-336.

DAWSON, A.G. (1994): Geomorphological effects of tsunami run-up and backwash.- Geomorphology, **10**: 83-94.

DAWSON, A.G. (1996): The geological significance of tsunamis.- Zeitschrift für Geomorphologie, N.F., **102**: 199-210.

DAWSON, A.G. (1999): Linking tsunami deposits, submarine slides and offshore earthquakes.- Quaternary International, **60**: 119-126.

DAWSON, A.G., LONG, D. & SMITH, D.E. (1988): The Storegga Slides: evidence from eastern Scotland for a possible tsunami.- Marine Geology, **82**: 271-276.

DAWSON, A.G., MUSSON, R.M.W., FOSTER, I.D.L. & BRUNSDEN, D. (2000): Abnormal historic sea-surface fluctuations, SW England.- Marine Geology, **170**: 59-68.

DAWSON, A.G., SMITH, D.E., RUFFMAN, A. & SHI, S. (1996): The diatom biostratigraphy of a tsunami sediment – examples from recent and middle Holocene events.- Physics and Chemistry of the Earth, **21** (12): 87-92.

DEBROT, A.O. & SYBESMA, J. (2000): The Dutch Antilles.- In: SHEPPARD, C.R.C. (ed.): Seas at the Millenium: An evironmental evaluation: 595-612; Pergamon, Amsterdam.

DE BUISONJÉ, P.H. (1964): Marine terraces and subaeric sediments on the Netherlands leeward Islands, Curaçao, Aruba and Bonaire, as indications of Quaternary changes in sea levels and climate.- Koninklijke Nederlandse Academie Wetenschapen, Proceedings Ser., **B 6**(1): 60-79.

DE BUISONJÉ, P.H. (1974): Neogene and Quaternary Geology of Aruba, Curaçao and Bonaire.- Natuurwetenschappelijke Studiekring Voor Surinam en de Nederlandse Antillen, **78**; Utrecht.

DE BUISONJÉ, P.H. & ZONNEVELD, J.I.S. (1960): De Kustvormen von Curaçao, Aruba en Bonaire.- Nieuwe West-Indische Gids, **40**: 121-144.

DE BUISONJÉ, P.H. & ZONNEVELD, J.I.S. (1976): Caracasbaai: A submarine slide of a huge coastal fragment in Curaçao.- Nieuwe West-Indische Gids, **51**: 55-88.

JONG DE, H. (1999): Wereldatlas voor de Nederlandse Antillen, Aruba en het Caribisch Gebied.- Hebri International B.V.; Amsterdam.

DE HAAN, D. & ZANEFELD, J.S. (1959): Some notes on tides in Annabaai Harbour Curaçao, Netherlands Antilles.- Bulletin of Marine Science, **9**: 224-236.

DE LANGE, W. & HEALY, T. (1999): Tsunami and tsunami hazard.- Tephra, **10**: 13-20.

DE VRIES, A.J. (2000): The semi-arid environment of Curaçao: a geochemical soil survey.- Geologie en Mijnbouw, **79**(4): 479-494.

DFG (2000): Meeresforschung im nächsten Jahrzehnt - Denkschrift.- Wiley-Verlag, Weinheim, XIII, 202 pp.

DOLAN, R. & DAVIS, R.E. (1994): Coastal Storm Hazards.- In: FINKL, CH.F. (ed.): Coastal Hazards. Perzeption, Susceptibility and Mitigation.- Journal of Coastal Research, Special Issue, **12**: 177-187; Fort Lauderdale.

DOMINEY-HOWES, D. (1996): The Geomorphology and Sedimentology of Five Tsunamis in the Aegean Sea Region, Greece.- Unpublished Ph.D. Thesis, University of Coventry, UK, 272 pp.

DONE, T.J. & DEVANTIER, L.M. (1986): Cyclone Winifred - observations on some ecological and geomophological effects. Workshop on the effects of cyclone Winifred.- Great Barrier Marine Park Authentic Workshop Series, **7**: 50-51.

DONE, T.J. & NAVIN, K.F. (1990): Shallow water benthic communities on coral reefs.- In: DONE, T.J. & NAVIN, K.F. (eds.): Vanuatu marine resources: report of a biological survey.- Australian Institute for Marine Science: 10-36; Townsville.

DONE, T.J. (1992): Effects of tropical cyclone waves on ecological and geomorphological structures on the Great Barrier Reef.- Continental Shelf Res., **12**: 859-872.

DRUFFEL, E.M. & LINICK, T.W. (1978): Radiocarbon in annual coral rings of Florida.- Geophysical Research Letters, **5**: 913-916.

DRUFFEL, E.M. (1982): Banded corals: changes in oceanic carbon-14 during the Little Ice Age.- Science, **218**: 13-19.

DRUFFEL, E.R.M. (1997): Pulses of rapid ventilation in the North Atlantic surface ocean during the past century.- Science, **275**: 1454-1457.

EXECUTIVE COUNCIL CURAÇAO (1995): Eilandelijke Ontwikkelingsplan Curaçao 1995.- (2 Vol.) DROV (Urban Planning and Housing Service); Curaçao.

FERNANDEZ, M., MOLINA, E., HASKOV, J. & ATAKAN, K. (2000): Tsunami and Tsunami Hazards in Central America.- Natural Hazards, **22**: 91-116.

FLETCHER, C.H., RICHMOND, B.M., BARNES, G.M. & SCHROEDER, T.A. (1995): Marine Flooding on the Coast of Kaua'i during Hurricane Iniki: Hindcasting Components and Delineating Washover.- Journal of Coastal Research., **11** (1): 188-204.

FOCKE, J.W. (1977): The effect of a potentially reef builing coralline algal community on an eroding limestone coast, Curaçao (Netherlands Antilles).- Proceedings 3rd International Coral Reef Symposium: 239-245; Miami.

FOCKE, J.W. (1978a): Geologische Aspekten van kwartaire Koraal- en Algenriffen op Curaçao, Bonaire, Aruba en Bermuda.- Ph.D. Thesis; Rijksuniversiteit te Leiden.

FOCKE, J.W. (1978b): Holocene development of coral fringing reefs, leeward of Curaçao and Bonaire (Netherlands Antilles).- Marine Geology, **28**: 31-41.

FOCKE, J.W. (1978c): Limestone cliff morphology and organism distribution on Curaçao (Netherlands Antilles).- Leidse Geologische Mededelingen, **51**(1): 131-150.

FOCKE, J.W. (1978d): Limestone cliff morphology on Curaçao (Netherlands Antilles), with special attention to the origin of notches and vermetid/coralline algal surf benches ("cornices","trottoirs").- Zeitschrift für Geomorphologie, N.F., **22**: 329-349.

FOCKE, J.W. (1978e): Subsea (0-40m) terraces and benches, windward of Curaçao, Netherlands Antilles.- Leidse Geologische Mededelingen, **51**(1): 95-102.

FOUKE, B. (1994): Deposition, Diagenesis and Dolomitization of Neogene Seroe Domi Formation Coral Reef Limestones on Curaçao, Netherlands Antilles.- Uitgaven Natuurwetenschappelijke Studiekring voor het Caraibisch Gebied, **134**: 182 pp.

FUKUI, Y. (1963): Hydrodynamic study on tsunami.- Coastal Engineering in Japan, **6**: 67-82.

GALAS, D. (1969): Die Calanquen der provencalischen Küste zwischen Cap Croisette und Cassis.- Geographische Rundschau, **11**: 420-423.

GENTRY, R.C. (1971): Hurricanes, one of the major features of airsea interaction in the Caribbean Sea.- Symposium of Investigations and Resources of the Caribbean Sea and Adjacent Regions, Curaçao, Nov. 18-26, 1968: 80-87.

GIERLOFF-EMDEN, H.G. (1980): Geographie des Meeres. Ozeane und Küsten.- 2 Vol.; Berlin, New York.

GRAY, W.M. (1990): Strong association between West African Rainfall and US Landfall of intense hurricanes.- Science, **249**: 1251-1256.

GRONTMIJ, S. & SOGREAH, M. (1968): Water and land resources development plan for the islands of Aruba, Bonaire and Curaçao.- 98 pp.; Wageningen.

GUINEY, J.L. (2000): Preliminary Report: Hurricane Lenny, 13-23 November 1999.- National Hurricane Center; URL: http://www.nhc.noaa.gov/1999lenny_text.html.

HAMILTON, R. (1941): Bijdrage tot de bodemkundige kennis van Nederlands West-Indie.- Tropengronde, **1**: 55 pp.

HARBITZ, C.B. (1991): Model simulations of tsunamis generated by the Storregga slide.- Institute of Mathematics, University of Oslo, Vol. **5**: 30 pp.

HARMELIN-VIVIEN, M.L. & LABOUTE, P. (1986): Catastrophic impact of atoll outer reef slopes in the Tuamotu (French Polynesia).- Coral Reefs, **5**: 55-62.

HARMELIN-VIVIEN, M. (1994): The Effects of Storms and Cyclones on Coral reefs: A Review.- In: FINKL, Ch.F. (ed.): Coastal Hazards. Perzeption, Susceptibility and Mitigation.- Journal of Coastal Research, Special Issue, **12**: 211-231; Fort Lauderdale.

HARTOG, J. (1968): Curaçao, from colonial dependance to autonomy.- History of the Netherlands Antilles, **3**: 424 pp.; De Wit, Aruba.

HAVISER, J.B. & JAY, B. (1987): Amerindian cultural geography on Curaçao.- Uitgaven Natuurwetenschappelijke Studiekring Voor Surinam en de Nederlandse Antillen, **120**: 212 pp.

HEARTY, J.P. (1997): Boulder deposits from large waves during the last interglaciation on North Eleuthera Island, Bahamas.- Quaternary Research, **48**(3): 326-338.

HECK, N.H. (1947): List of seismic seawaves.- Bulletin of the Seismological Society of America, **37** (4): 269-286.

HELMERS, H. & BEETS, D.J. (1977): Geology of the Cretaceous of Aruba.- Field Guide 8th Caribbean Geological Conference (Curaçao), **15**: 29-35.

HERBRICH, J.B. (1990): Wave run-up and overtopping.- In: Handbook of Coastal Engineering.- Vol. **1**; Gulf Publishing Co., Houston.

HERNANDEZ-AVILA, M.L., ROBERTS, H.H. & ROUSE, L.J. (1977): Hurricane-generated waves and boulder rampart formation.- 3rd International Coral Reef Symposium: 71-78.

HERWEIJER, J.P., DE BUISONJÉ, P.H. & ZONNEVELD, J.I.S. (1977): Neogene and Quaternary geology and geomorphology.- 8th Caribbean Geological Conference.

HERWEIJER, J.P. & FOCKE, J.W. (1978): Late Pleistocene depositional and denudational history of Aruba, Bonaire and Curaçao (Netherlands Antilles).- Geologie en Mijnbouw, **57**2): 177-187.

HILLS, J.C. & GODA, P. (1998): Tsunami from Asteroid and Comet Impacts: The Vulnerability of Europe.- Science of Tsunami Hazards, **16**(1): 3-10.

HITTELMANN, A.M., LOCKRIDGE, P.A., WHITESIDE, L.S. & LANDER, J.F. (2001): Interpretive Pitfalls in Historical Hazards Data.- Natural Hazards, **23**: 315-338.

HODKIN, E.P. (1964): Rate of erosion of intertidal limestones.- Zeitschrift für Geomorphologie, N.F., **8**: 385-392.

HORIKAWA, K. & SHUTO, N. (1983): Tsunami Disaster and Protection Measures in Japan.- Proceedings International Tsunami Symposium 1981, IUGG Tsunami Commission, May 1981; Sendai-Ofunato-Kamaishi, Japan; Terra Publishing Co..

HUGHEN, K.A., OVERPECK, J.T., PETERSON, L.C. & ANDERSON, R.F. (1996): The nature of varved sedimentation in the Cariaco Basin, Venezuela, and its palaeoclimatic significance.- In: KEMP, A.E.S. (ed.): Palaeoclimatology and Palaeoceanography from Laminated Sediments: 171-183.

HUGHEN, K., SOUTHON, J.K., LEHMAN, S. & OVERPECK, J. (2000): Synchronous Radiocarbon and Climate Shift During the Last Deglaciation.- Science, **290**: 1951-1954.

HUTCHINSON, P.A. (1986): Biological destruction of coral reefs.- Coral Reefs, **4**: 239-252.

IIDA, K., COX, D.C. & PARARAS-CARAYANNIS, G. (1967): Preliminary catalogue of tsunamis occurring in the Pacific Ocean.- Data Report Nr. **5**: 274 pp.; University of Hawaii, Honolulu.

JARVINEN, B.R., NEUMANN, C.J. & DAVIS, M.A.S. (1984): A tropical cyclone data tape for the North Atlantic Basin, 1886-1983: Contents, limitations, and uses.- NWS NHC 22: 21p; Coral Gables, Florida.

JOHNSON, M.E. (1997): Caribbean Storm Surge Return Periods: Final Report.- URL: http://www.oas.org/EN/cdmp/document/johnson/statpapr.htm.

JONES, B. & HUNTER, I.G. (1992): Very Large Boulders on the Coast of Grand Cayman: the Effects of Giant Waves on Rocky Shorelines.- Journal of Coastal Research, **8**(4): 763-774.

KAWANA, T. & PIRAZZOLI, P.A. (1990): Re-examination of the Holocene emerged shorelines in Irabu and Shimoji Islands, the South Ryukyus, Japan.- Quaternary Research, **28**: 419-426.

KELLETAT, D. (1985): Biodestruktive und biokonstruktive Formenelemente an der spanischen Mittelmeerküste.- Geoökodynamik, **VI**(1/2): 1-20.

KELLETAT, D. (1989): Main Aspects of Coastal Evolution of Sylt (FRG).- In: BIRD, E.C.F. & KELLETAT, D. (eds.): Zonality of Coastal Geomorphology and Ecology.- Program, Abstracts and Field Guide of the Pre-Conference Symposium, 2nd International Conference on Geomorphology in Frankfurt: 25-53.

KELLETAT, D. (1998): Geologische Belege katastrophaler Erdkrustenbewegungen 365 AD im Raum von Kreta.- In: OLSHAUSEN, E. & SONNABEND, H. (eds.): Naturkatastrophen in der antiken Welt.- Geographica Historica, **10**: 156-161; Stuttgarter Kolloquium zur Historischen Geographie des Altertums, Stuttgart.

KELLETAT, D. (1999): Physische Geographie der Meere und Küsten.- 2nd ed.; 258 pp.; Stuttgart, Teubner.

KELLETAT, D. & SCHELLMANN, G. (2001a): Sedimentologische und geomorphologische Belege starker Tsunami-Ereignisse jung-historischer Zeitstellung im Westen und Südosten Zyperns.- Essener Geographische Arbeiten, **32**: 1-74.

KELLETAT, D. & SCHELLMANN, G. (2001b): Tsunamis in Cyprus: Field Evidences and ^{14}C Dating Results.- Zeitschrift für Geomorphologie, NF; (*in press*).

KEULEGAN, G.H. (1950): Engineering Hydraulics.- New York (Wiley), 768 pp.

KJERFVE, B., MAGILL, K.E., PORTER, J.W. & WOODLEY, J.D. (1986): Hindcasting of hurricane characteristics and observed storm damage on a fringing reef, Jamaica, West Indies.- Journal of Marine Research, **44**: 119-148.

KLAVER, G.T. (1987): The Curaçao Lava Formation: an ophiolitic analogue of the anomalous thick layer 2B of the Mid-Cretaceous oceanic plateaus in the Western Pacific and Central Caribbean.- GUA Papers of Geology, **27**: 1-68.

KLUG, H. (1986): Flutwellen und Risiken der Küste.- 122 pp.; Steiner, Wiesbaden.

KOBLUK, D.R. & LYSENKO, M.A. (1992): Storm Features on a Southern Caribbean Fringing Coral Reef.- Palaios, **7**: 213-221.

KOHN, A.J. (1980): Populations of tropical intertidal gastropods before and after a typhoon.- Micronesia, **16**: 215-228.

KRIVOSHEY, M.I. (1970): Experimental investigations of tsunami waves.- In: ADAMS, W.M. (ed.): Tsunamis in the Pacific Ocean: 351-365; East-West Press Center, Honolulu.

LANDER, J.F. (1996): Tsunamis Affecting Alaska 1737-1996.- NGDC Key to Geophysical Record Documentation No. **31**; 195 pp.; NOAA.

LANDER, J.F. & LOCKRIDGE, P.A. (1989): United States Tsunamis (including United States Possessions) 1690-1988.- Vol. **41/42**; NOAA/National Geophysical Data Center, Boulder.

LANDER, J.F., LOCKRIDGE, P.A. & KOZUCH, M.J. (1993): Tsunamis affecting the West Coast of the United States, 1806-1992; United States Department of Commerce, Boulder, Colorado.

LANDER, J.F., O'LOUGHLIN, K.F. & WHITESIDE, L.S. (1998): Caribbean Tsunamis: A 500-Year History, 1498 to 1998.- Natural Hazards (submitted).

LANDER, J.F. & WHITESIDE, L.S. (1997): Caribbean Tsunamis: An Initial History.- Tsunami Workshop June 11-13, Mayaguez, Puerto Rico. URL: http://www.cima.-uprm.edu/~tsunami/Lander/J_Lander.html.

LANDSEA, C.W. (1993): A climatology of intense (or major) Atlantic hurricanes.- Monthly Weather Review, 121: 1703-1713.

LANDSEA, C.W. (1998): Climate Variability of Tropical Cyclones: Past, Present and Future.- In: PIELKE, R.A. (Sr.) & PIELKE, R.A. (Jr.) (eds.): Storms: 220-241.

LANDSEA, C.W., PIELKE, R.A., MESTAS-NUNEZ, A.M. & KNAFF, J.A. (1999): Atlantic Basin Hurricanes: Indices of Climatic Change.- Climatic Change, 42: 89-129.

LATTER, J.H. (1981): Tsunami of Volcanic Origin: Summary of Causes, with Particular Reference to Krakatoa, 1883.- Bulletin Volcanologique, 44(3): 467-490.

LIGHTY, R., MACINTYRE, I. & STUCKENRATH, R. (1982): Acropora palmata reef framework: A reliable indicator of sea level in the western Atlantic for the past 10,000 years.- Coral Reefs, 1: 125-130.

LIPMAN, P.W., NORMARK, W.R., MOORE, J.G., WILSON, J.B. & GUTMACHER, C. (1988): The giant submarine Alika Debris Slide, Mauna Loa, Hawaii.- Journal of Geophysical Research, 93: 4279-4299.

LOCKRIDGE, P.A. (1988): Historical Tsunamis in the Pacific Basin.- In: EL-SABH, M.I. & MURTY, T.S. (eds.): Natural and Man-Made Hazards: 171-181; Reidel Publishing, Dordrecht.

LORANG, M.S. (2000): Predicting Threshold Entrainment Mass for a Boulder Beach.- Journal of Coastal Research, 16(2): 432-445.

LOWE, D.J. & DE LANGE, W.P. (2000): Volcano-meteorological tsunamis, the AD 200 Taupo eruption (New Zealand) and the possibility of a global tsunami.- The Holocene, 10(3): 401-407.

LUDLUM, D.M. (1989): Early American Hurricanes, 1492-1870.- 198 pp.; Lancaster Press, Pennsylvania.

MACDONALD, R., HAWKESWORTH, C.J. & HEATH, E. (2000): The Lesser Antilles Volcanic Chain: a study in arc magmatism.- Earth Science Reviews, 69: 1-79.

MACKEE (1959): Storm sediments on a Pacific Atoll.- Journal of Sedimentary Petrology, 29: 354-364.

MADER, C.L. (1996): Asteroid Tsunami Inundation of Hawaii.- Science of Tsunami Hazards, 14(2): 85-88.

MADER, C.L. (1998): Modeling the Eltanin Asteroid Tsunami.- Science of Tsunami Hazards, 16(1): 17-20.

MALONEY, N.J. (1967): Geomorphology of the continental margin of Venezuela. Part 3. Bonaire Basin (66°W to 70°W longitude).- Bolivia Institute of Oceanography, 6: 286-302.

MANN, P., SCHUBERT, C. & BURKE, K. (1990): Review of Caribbean Neotectonics.- In: DENGO, G. & CASE, J.E. (eds.): The Caribbean Region. The Geology of North America, Vol. 2: 207-238; Geological Society of America, Boulder.

MARAGOS, J.E., BAINES, G.B.K. & BEVERIDGE, J.P. (1973): Tropical Cyclone Bebe creates a new land formation on Funafuti Atoll.- Science, 181: 1161-1164.

MARTIN, K. (1888): Bericht über eine Reise nach Niederländisch West-Indien und darauf gegründete Studien.- 238 pp.; Leiden.

MASSEL, S.R. (1989): Hydrodynamics of Coastal Zones.- 336 pp.; Elsevier.

MASSEL, S.R. & DONE, J.T. (1993): Effects of cyclone waves on massive coral assemblages on the Great Barrier Reef: meteorology, hydrodynamics and demography.- Coral Reefs, 12: 153-166.

MASTRONUZZI, G. & SANSO, P. (2000): Boulder transport by catastrophic waves along the Ionian coast of Apulia (southern Italy).- Marine Geology, 170: 93-103.

MAUL, G.A. (1999): On the role of IOCARIBE in a Caribbean Tsunami System: Science, Engineering, Management and Education.- Marine Geodesy Journal, 22: 1-13.

MAUL, G.A., HENDRY, M.D. & PIRAZOLLI, P.A. (1996): Sea Level, Tides, and Tsunamis.- Small Islands: Marine Science and Sustainable Development. Coastal and Estuarine Studies, 51: 83-119.

MCLEAN, R.F. (1992): A two thousand year history of low latitude tropical storms. Prelimininary results for Funafuti, Tuvalu.- 7th International Coral Reef Symposium, abstract: 66 p.

MCSAVENEY, M.F., GOFF, J.R., DARBY, D.J., GOLDSMITH, P., BARNETT, A., ELLIOTT, S. & NONGKAS, M. (2000): The 17 July 1998 tsunami, Papua New Guinea: evidence and initial interpretation.- Marine Geology, 170: 81-92.

METEOROLOGICAL SERVICE OF THE NETHERLANDS ANTILLES AND ARUBA (1998): Hurricanes and Tropical Storms of the Netherlands Antilles and Aruba.- URL: http://www.meteo.an.

METEOROLOGICAL SERVICE OF THE NETHERLANDS ANTILLES AND ARUBA (2001): Climate Reports.- URL: http://www.meteo.an.

MILLÁS, J.C. (1968): Hurricanes of the Caribbean and adjacent regions, 1492-1800.- 328 pp.; Academy of Arts and Sciences of the Americas, Miami.

MOLNAR, P. & SYKES, L.R. (1969): Tectonics of the Caribbean and Middle America Regions from Focal Mechanisms and Seismicity.- Geological Society of America Bulletin, 80: 1639-1684.

MOORE, G.W. & MOORE, J.G. (1984): Deposits from a giant wave on the island of Lanai, Hawaii.- Science, 226: 1312-1315.

MOORE, G.W. & MOORE, J.G. (1988): Large-scale bedforms in boulder gravel produced by giant waves in

Hawaii.- Geological Society of America, Special Issue, **229**: 101-110.

MOORE, J.G., NORMARK, W.R. & GUTMACHER, C.E. (1992): Major landslides on the submarine flanks of the Mauna Loa Volcano, Hawaii.- Landslide News, **6**: 13-16.

MOYA, J.C. (1999): Stratigraphical and morphologic evidence of tsunami in northwestern Puerto Rico.- Submitted to Sea Grant College Program, University of Puerto Rico, Mayaguez Campus.

MUESSIG, K.W. (1984): Structure and Cenozoic tectonics of the Falcon Basin, Venezuela, and adjacent regions.- In: BONINI, W.E., HARGRAVES, R.B. & SHAGAM, R. (eds.): The Caribbean South-American Plate Boundary and Regional Tectonics: 217-231.

NATIONAL GEOPHYSICAL DATA CENTER (2001): Tsunami Data at NGDC.- URL:http://www.ngdc.noaa.gov/seg/hazard/tsu.shtml.

MURTY, T.S. (1977): Seismic Sea Waves – Tsunamis.- Dep. of Fisheries and the Environment, Fisheries and Marine Service; Ottawa, Canada.

NAKATA, T. & KAWANA, T. (1993): Historical and prehistorical large tsunamis in the southern Ryukyus, Japan.- "Tsunamis '93": 297-307; Wakayama, Japan,

NEUMANN, C.J., JARVINEN, B.R., MCADIE, C.J. & ELMS, J.D. (1993): Tropical cyclones of the North Atlantic Ocean, 1871-1992.- National Climatic Data Center in cooperation with the National Hurricane Center: 193 pp.; Coral Gables, Florida.

NEUMANN, C.J. & MCADIE, C.J. (1997): The Atlantic tropical cyclone file: A critical need for a revision.- 22th Conference on Hurricanes and Tropical Meteorology: 401-402; American Meteorological Society, Boston.

NEWMAN, J.N. (1977): Marine Hydrodynamics.- 402 pp.; MIT Press, Cambridge, London.

NISBET, E.G. & PIPER, D.J.W. (1998): Giant submarine landslides.- Nature, **392**: 329-330.

NOTT, J. (1997): Extremely high wave deposits inside the Great Barrier Reef, Australia: determining the cause - tsunami or tropical cyclone.- Marine Geology, **141**: 193-207.

NOTT, J. (2000): Records of Prehistoric Tsunamis from Boulder Deposits – Evidence from Australia.- Science of Tsunami Hazards, **18**(1): 3-14.

OAK, H.L. (1985): Process inference from coastal-protection structures of boulder beaches.- Geografiska Annaler, **68 A**: 25-31.

OGG, J.G. & KOSLOW, J.A. (1978): The impact of Typhoon Pamela (1976) on Guam's coral reefs and beaches.- Pacific Science, **32**: 105-118.

OTA, Y., PIRAZZOLI, P.A., KAWANA, T. & MORIWAKI, H. (1985): Late Holocene coastal geomorphology and sea-level records on three small islands, the South Ryukyus, Japan.- Geographical Review Japan, **58 B**: 185-194.

PANDOLFI, J.M., LLEWELLYN, G. & JACKSON, J.B.C. (1999): Pleistocene reef environments, constituent grains and coral community structure: Curaçao, Netherlands Antilles.- Coral Reefs, **18**: 107-122.

PAPADOPOULOS, G.A. & CHALKIS, B.J. (1984): Tsunamis observed in Greece and the surrounding area from antiquity up to present times.- Marine Geology, **56**: 309-317.

PARARAS-CARAYANNIS, G. (1969): Catalog of Tsunamis in the Hawaiin Islands.- Report WDCA-T **69-2**: 94 pp.; ESSA - Coast and Geodetic Survey, Boulder, Colorado.

PARARAS-CARAYANNIS, G. (1983): The Tsunami Impact on Society.- Proceedings International Tsunami Symposium 1981, IUGG Tsunami Commission, May 1981: 3-8; Sendai-Ofunato-Kamaishi, Japan, Terra Publishing Co.

PERKINS, R. & ENOS, P. (1968): Hurricane Betsy in the Florida-Bahama area – geological effects and comparison with hurricane Donna.- Journal of Geology, **76**: 710-717.

PINDALL, J.L. & BARETT, S.F. (1990): Geological evolution of the Caribbean region: a plate tectonic perspective.- In: DENGO, G. & CASE, J.E. (eds.): The Geology of North America: 405-432; Boulder.

PIRAZZOLI, P.A., MONTAGGIONI, L.F., SALVAT, B. & FAURE, G. (1988): Late Holocene sea level indicators from twelve atolls in the central and eastern Tuamotus (Pacific Ocean).- Coral Reefs, **7**: 57-68.

POSER, H. & HÖVERMANN, J. (1952): Beiträge zur morphometrischen und morphologischen Schotteranalyse.- Abhandl. Braunschweigische Wissenschaftliche Gesellschaft, **IV**: 12-36.

RADTKE, U.,SCHELLMANN, G., SCHEFFERS, A., KELLETAT, D., KASPARS, H.U. & KROMER, B. (2002): ESR and ^{14}C dating of corals deposited by Holocene tsunami events on the Netherlands Antilles (Curaçao, Bonaire, Aruba).- Quaternary Science Reviews (*in press*).

READING, A.J. (1989): Caribbean tropical storm activity over the past few centuries.- International Journal of Climatology, **10**: 365-376.

REED, J.K. (2001): Submersible and Scuba Collections in the coastal waters of the Netherlands Antilles (Curaçao, Bonaire) and Aruba: Biomedical and Biodiversity Research of the benthic communities with emphasis on the Porifera and Gorgonacea.- Harbor Branch Oceanographic Institution, Inc., Ft. Pierce, Florida.

ROBSON, G.R. (1964): An earthquake catalogue for the Eastern Caribbean, 1530-1960.- Bulletin of the Seismological Society of America, **54**: 785-832.

ROJER-BEENHAKKER, A.C. (1985): Een bijdrage tot de vegetatie-analyse van de Curaçaose kalksteengebieden.- , 54 pp.; Unpublished Report Institut Bijzondere Plantkunde, Utrecht University.

RULL, V. (2000): Holocene sea level rising in Venezuela: a

preliminary curve.- Boletín de la Sociedad Venezolana de Geólogos *(in press)*; URL: http://www.ecopal.org/sealevel.htm.

SARMIENTO, G. (1976): Evolution of arid vegetation in tropical america.- In: GOODALL, E.W. (ed.): Evolution of desert biota: 65-99.

SARPKAYA, T. & ISAACSON, M. (1981): Mechanics of Wave Forces on Offshore Structures.- 651pp.; Van Nostrand Reinold Co.

SCHEFFERS, A. & KELLETAT, D. (2001): Hurricanes und Tsunamis: Dynamik und küstengestaltende Wirkungen.- Bamberger Geographische Schriften, **20**: 29-53.

SCHMIDT, W. (1923): Die Scherms an der Rotmeerküste von el-Hedschas.- Petermanns Geogr. Mitteilungen, **69**: 118-121.

SCHUBERT, C. & VALASTRO, S. (1976): Quaternary geology of La Orchila Island, central Venezuelan offshore, Caribbean Sea.- Geological Society of America Bulletin, **87**: 1131-1142.

SCHUBERT, C. (1988): Neotectonics of the La Victoria Fault Zone, North-Central Venezuela.- Annales Tectonicae, **II**: 58-66.

SCHUBERT, C. (1994): Tsunamis in Venezuela: Some Observations on their Occurrence.- In: FINKL, CH.F. (ed.): Coastal Hazards. Perception, Susceptibility and Mitigation.- Journal of Coastal Research, Special Issue **12**: 189-195; Fort Lauderdale.

SCHÜLKE, H. (1969): Bestimmungsversuch des Rias-Begriffes durch das Kriterium der Fluvialität (mit einem Ausblick auf das Ästuarproblem).- Erdkunde, **23**: 264-280.

SCOFFIN, T.P. (1993): The geological effects of hurricanes on coral reefs and the interpretation of storm deposits.- Coral Reefs, **12**: 203-221.

SIMKIN, T. & FISKE, R.S. (1983): Krakatau 1883: the volcanic eruption and its effects.- 464 pp.; Smithsonian Institution Press, Washington DC.

SIMPSON, R.H. & RiEhl, H. (1981): The hurricane and its impact.- 398 pp.; Louisiana State University Press, Baton Rouge.

SINTON, C.W., DUNCAN, R.A. & STOREY, M. (1993): ^{40}Ar/^{39}Ar ages from Gorgona Island, Colombia and the Nicoya Peninsula, Costa Rica.- EOS, **74**: abstract.

SINTON, C.W., DUNCAN, R.A., STOREY, M., LEWIS, J. & ESTRADA, J.J. (1998): An oceanic flood basalt province within the Caribbean plate.- Earth and Planetary Science Letters, **155**: 221-235.

SMITH, M.S. & SHEPHERD, J.B. (1994): Explosive Submarine Eruptions of Kick 'em Jenny Volcano: Preliminary Investigations of the Potential Tsunami Hazard in the Eastern Caribbean Region.- In: AMBEH, W.B. (ed.): Proceedings, Caribbean Conference on Natural Hazards: Volcanoes, Earthquakes, Windstorms, Floods: 249-260; University of the West Indies, Mona Campus.

SMITH, M.S. & SHEPHERD, J.B. (1996): Tsunami waves generated by volcanic landslides: an assessment of the hazard associated with Kick 'em Jenny.- In: MCGUIRE, W.J., JONES, A.P. & NEUBERG, J. (eds.): Volcano instability on earth and other planets: 115-123.

SOLOVIEV, S.L. (1970): Recurrence of tsunamis in the Pacific.- In: ADAMS, W.M. (ed.): Tsunamis in the Pacific Ocean: 149-164; University of Hawaii, Honolulu.

SOLOVIEV, S.L. & GO, C.N. (1974a): Catalogue of Tsunamis on the Western shore of the Pacific Ocean (1736-1968).- 310 pp.; Canadian Translations of Fisheries and Aquatic Sciences, **5077**; Nauka Publishing House, Moscow.

SOLOVIEV, S.L. & GO, C.N. (1974b): Catalogue of Tsunamis on the Eastern shore of the Pacific Ocean (1513-1968).- 204 pp.; Canadian Translations of Fisheries and Aquatic Sciences, **5078**; Nauka Publishing House, Moscow.

SOLOVIEV, S.L., GO, C.N. & KIM, K.S. (1992): Catalogue of Tsunamis in the Pacific 1969-1982.- 208 pp.; Academy of Science, Moscow.

SPENNEMANN, D.H. & HEAD, M.J. (1996): Reservoir modification of radiocarbon signatures in coastal and near-shore waters of Eastern Australia: the state of the play.- URL: http://life.csu.edu.au/~dspennem/VIRTPAST/C14/OCR/OCR_OZ_Review96.html.

STEPHENSON, W., ENDEAN, R. & BENNETT, I. (1958): An ecological survey of the marine fauna of Low Isles, Queensland.- Australian Journal of Marine and Freshwater Research, **9**: 261-318.

STIENSTRA, P. (1983): Quaternary sea-level fluctuations on the Netherlands Antilles – possible correlations between a newly composed sea-level curve and local features.- Marine Geology, **52**(1-2): 27-37.

STIENSTRA, P. (1991): Sedimentary Petrology, Origin and Mining History of the Phosphate Rocks of Klein-Curaçao, Curaçao and Aruba, Netherlands West Indies.- Publications Foundation for Scientific Research in the Caribbean Region, **130**: 207 pp.

STODDART, D.R. (1962): Catastrophic storm effects on the British Honduras reefs and cays.- Nature, **196**: 512-515.

STODDART, D.R. (1963): Effects of Hurricane Hattie on the British Honduras reefs and cays, October 30-31, 1961.- Atoll Research Bulletin, **95**: 1-142.

STODDART, D.R. (1971): Coral reefs and islands and catastrophic storms.- In: STEERS, J.A. (ed.): Applied Coastal Geomorphology: 155-197; Macmillan Company, London.

STODDART, D.R., MCLEAN, R.F., SCOFFIN, T.P. & GIBBS, P.E. (1978): Forty-five years of change on low wooded islands, Great Barrier Reef.- Philosophical Transactions of the Royal Society of London, Ser. B, **284**: (999): 63-80.

STOFFERS, A.L. (1956): The vegetation of the Netherlands Antilles.- Publication of the Foundation Scientific Research Surinam Neth. Ant., **25**: 142 pp.; Utrecht.

STOFFERS, A.L. (1980): Flora and vegetation of the Leeward Islands I: General Introduction and coastal communities.- Miscellaneous Papers Landbouwhogesch. Wageningen, **19**: 293-314.

STUIVER, M., PEARSON, G.W. & BRAZIUNAS, T.F. (1986): Radiocarbon age calibration of marine samples back to 9000 cal yr BP.- Radiocarbon, **28**: 980-1021.

STUIVER, M. & BRAZIUNAS, T.F. (1993): Modeling atmospheric ^{14}C influences and ^{14}C ages of marine samples to 10,000 BC.- Radiocarbon, **35**: 137-189.

STUIVER, M., REIMER, P. & BRAZIUNAS, T.F. (1998): High-precision radiocarbon age calibration for terrestrial and marine samples.- Radiocarbon, **40**: 1127-1151.

SYKES, L.R., MCCANN, W.R. & KAFKA, A.L. (1982): Motion of the Caribbean plate during the last 7 million years and implications for earlier Cenozoic movements.- Journal of Geophysical Research, **87**: 10656-10676.

TEENSTRA, M.D. (1977): De Nederlandse West-Indische eilanden in derzelve tegenwoordigen toestand 1836-1937. I. Curaçao - II. Curaçao, St. Maarten, St. Eustatius, Saba.- 381 pp.; Sulpke, Amsterdam.

TEISSIER, R. (1969): Les Cyclones en Polynesie Francaise.- Bulletin Société Etudes Oceaniennes, **14**: 20-21.

TERPSTRA, H. (1948): De boomgroei op de Benedenwindse eilanden in vroeger tijd.- Mededelingen Koning. Vereining Indisch Inst., **LXXVIII**: 3-19.

TOMBLIN, J. (1981): Earthquakes, volcanoes and hurricanes. A review of natural hazards and vulnerability in the West Indies.- Ambio, **10**: 340-345.

TRUDGILL, S.T. (1976): The marine erosion of limestone on Aldabra Atoll, Indian Ocean.- Zeitschrift für Geomorphologie, N.F., **26**: 164-200.

TRUDGILL, S.T. (1983): Measurements of rates of erosion of reefs and reef limestones.- In: BARNES, D.J. (ed.): Perspectives on coral reefs: 256-262; Australian Institute of Marine Science, Townsville.

UNESCO (1991): Tsunami Glossary. A Glossary of Terms and Acronyms Used in the Tsunami Literature.- Technical Series, **37**.

VAN DE HOEK, C. (1969): Algal vegetation types along the open coasts of Curaçao, Netherlands Antilles.- Koninklijke Nederlandse Academie Wetenschapen, Proceedings, **C 72**: 537-577.

VAN DUYL, F.C. (1985): Atlas to the living reefs of Curaçao and Bonaire (Netherlands Antilles).- 39 pp.; Utrecht.

VAN LOENHOUD, P.J. & VAN DE SANDE, J.C.P.M. (1977): Rocky shore zonation on Aruba and Curaçao (Netherlands Antilles) with the introduction of a new general scheme of zonation.- Koninklijke Nederlandse Academie Wetenschapen, Proceedings, **C 80/1**: 437-474.

VAN MOORSEL, G.W.N.M. & MEIJER, A.J.M. (1993): Base-Line Ecological Study van het Lac op Bonaire.- 120 pp.; Bureau waardenburg BV, The Netherlands.

VAN WOESIK, R., AYLING, A.M. & MAPSTONE, B. (1991): Impact of tropical cyclone Ivor on the Great Barrier Reef, Australia.- Journal of Coastal Research, **7**: 551-558.

WATANABE, H. (1998): Catalogue of Hazardous Tsunamis in Japan.- 238 pp.; 2nd version (in Japanese), University of Tokyo Press.

WEISS, M.P. (1979): A saline lagoon on Cayo Sal, western Venezuela.- Atoll Research Bulletin, **232**: 1-33.

WESTERMANN, J.H. (1932): The geology of Aruba.- Ph.D. Thesis, Geografische Geologische Mededelingen Physiographie, **7**: 133 pp.; Utrecht.

WHITE, R.V., TARNEY, J., KERR, A.C., SAUNDERS, A.D., KEMPTON, P.D., PRINGLE, M.S. & KLAVER, G.T. (1999): Modification of an oceanic plateau, Aruba, Dutch Caribbean: Implications for the generation of continental crust.- Lithos, **46**: 43-68.

WIEGEL, R.L. (1976): Tsunamis.- In: Comnitz, C. & Rosenblueth, E. (eds.): Seismic Risk and Engineering Decisions.- Elsevier, Amsterdam.

WOODLEY, J.D. & et.al. (1981): Hurricane Allen's impact on Jamaican coral reefs.- Science, **214**: 749-755.

WOODLEY, J.D. (1990): The effects of Hurricane Gilbert on coral reefs at Discovery Bay.- In: BACON, F.R. (ed.): Assessment of the Economic impacts of Hurricane Gilbert on Coastal and Marine Resources in Jamaica.- 71-73.

YOUNG, R.W. & BRYANT, B. (1992): Catastrophic wave erosion on the southeastern coast of Australia: impact of the Lanai tsunami ca.105 ka?- Geology, **20**: 199-202.

ZHOU, Q. & ADAMS, W.M. (1986): Tsunamigenic earthquakes in China: 1831 BC to 1980 AD.- Science of Tsunami Hazards, **4(3)**: 131-148.

ESSENER GEOGRAPHISCHE ARBEITEN

Band 1: Ergebnisse aktueller geographischer Forschungen an der Universität Essen. - 207 Seiten, 47 Abbildungen, 30 Tabellen. 1982. • 8.- (ISBN 3-506-72301-4)

Band 2: G. HENKEL (Hg.): Dorfbewohner und Dorfentwicklung. Vorträge und Ergebnisse der Tagung in Bleiwäsche vom 17.-19. März 1982. - 127 Seiten, 6 Abbildungen. 1982. • 5.- (ISBN 3-506-72302-3)

Band 3: J.-F. VENZKE: Geoökologische Charakteristik der wüstenhaften Gebiete Islands. - 206 Seiten, 44 Abbildungen, 15 Tabellen, 1 Karte als Beilage. 1982. • 9.- (ISBN 3-506-72303-0)

Band 4: J. BIEKER & G. HENKEL: Erhaltung und Erneuerung auf dem Lande. Das Beispiel Hallenberg. - 255 Seiten, 53 Abbildungen, 46 Tabellen, 55 Schwarz-Weiß-Fotos, 2 Farbfotos. 1983. • 13.- (ISBN 3-506-72304-9)

Band 5: W. TRAUTMANN: Der kolonialzeitliche Wandel der Kulturlandschaft in Tlaxcala. Ein Beitrag zur historischen Landeskunde Mexikos unter besonderer Berücksichtigung wirtschafts- und sozialgeographischer Aspekte. - 420 Seiten, 13 Figuren, 25 Fotos, 10 Karten, 7 Tabellen. 1983. • 13.- (ISBN 3-506-72305-7)

Band 6: D. KELLETAT (Hg.): Beiträge zum 1. Essener Symposium zur Küstenforschung. - 312 Seiten, 97 Abbildungen, 7 Tabellen. 1983. • 13.- (ISBN 3-506-72306-5)

Band 7: D. KELLETAT: Internationale Bibliographie zur regionalen und allgemeinen Küstenmorphologie (ab 1960). - 218 Seiten, 1 Abbildung. 1983. • 8.- (ISBN 3-506-72307-3)

Band 8: G. HENKEL & H.-J. NITZ (Hg.): Ländliche Siedlungen einheimischer Völker Außereuropas – Genetische Schichtung und gegenwärtige Entwicklungsprozesse. Arbeitskreissitzung des 44. Deutschen Geographentages. - 148 Seiten, 35 Abbildungen, 21 Fotos, 4 Karten, 5 Tabellen. 1984. • 8.- (ISBN 3-506-72308-1)

Band 9: H.-W. WEHLING: Wohnstandorte und Wohnumfeldprobleme in der Kernzone des Ruhrgebietes. - 285 Seiten, 38 Abbildungen, 24 Tabellen, 10 Übersichten. 1984. • 13.- (ISBN 3-506-72309-X)

Band 10: D. KELLETAT (Hg.): Beiträge zur Geomorphologie der Varanger-Halbinsel, Nord-Norwegen (KELLETAT: Studien zur spät- und postglazialen Küstenentwicklung der Varanger-Halbinsel, Nord-Norwegen; MEIER: Studien zur Verbreitung, Morphologie, Morphodynamik und Ökologie von Palsas auf der Varanger-Halbinsel, Nord-Norwegen). - 243 Seiten, 63 Abbildungen, 8 Figuren, 45 Tabellen. 1985. • 13.- (ISBN 3-506-72310-3)

Band 11: D. KELLETAT: Internationale Bibliographie zur regionalen und allgemeinen Küstenmorphologie (ab 1960). - 1. Supplementband (1960-1985). 244 Seiten, 1 Abbildung. 1985. • 8.- (ISBN 3-506-72311-1)

Band 12: H.-W. WEHLING: Das Nutzungsgefüge der Essener Innenstadt. - Bestand und Veränderungen seit 1978. - 1986 (ISBN 3-506-72312-X) *vergriffen*

Band 13: W. KREUER: Landschaftsbewertung und Erholungsverkehr im Reichswald bei Kleve. Eine Studie zur Praxis der Erholungsplanung. - 205 Seiten, 45 Abbildungen, 46 Tabellen. 1986. • 10.- (ISBN 3-506-72313-8)

Band 14: D. KELLETAT & H.-W. WEHLING: Beiträge zur Geographie Nord-Schottlands (KELLETAT: Die Bedeutung biogener Formung im Felslitoral Nord-Schottlands. WEHLING: Leben am Rande Europas. Wirtschafts- und Sozialstrukturen in der Crofting-Gemeinde Durness). - 176 Seiten, 85 Abbildungen, 6 Tabellen. 1986. • 10.- (ISBN 3-506-72314-6)

Band 15: G. HENKEL (Hg.): Kommunale Gebietsreform und Autonomie im ländlichen Raum. Vorträge und Ergebnisse der Tagung in Bleiwäsche vom 12.-13. Mai 1986. - 160 Seiten. 1986. • 8.- (ISBN 3-506-72315-4)

Band 16: G. HENKEL (Hg.): Kultur auf dem Lande. Vorträge und Ergebnisse des 6. Dorfsymposiums in Bleiwäsche vom 16.-17. Mai 1988. - 231 Seiten, 22 Fotos, 2 Tabellen. 1988. • 11.- (ISBN 3-506-72316-4)

Band 17: D. KELLETAT (Hg.): Neue Ergebnisse zur Küstenforschung. Vorträge der Jahrestagung Wilhelmshaven 18. und 19. Mai 1989. - 388 Seiten, 23 Fotos, 119 Abbildungen. 1989. • 13.- (ISBN 3-506-72317-0)

Band 18: E. C. F. BIRD & D. KELLETAT (Eds.): Zonality of Coastal Geomorphology and Ecology. Proceedings of the Sylt Symposium, August 30 - September 3, 1989. - 295 Seiten, 88 Fotos, 52 Abbildungen, 7 Tabellen. 1989. • 15.- (ISBN 3-506-72318-9)

Band 19: G. HENKEL & R. TIGGEMANN (Hg.): Kommunale Gebietsreform - Bilanzen und Bewertungen. Beiträge und Ergebnisse der Fachsitzung des 47. Deutschen Geographentages Saarbrücken 1989. - 124 Seiten, 9 Abbildungen, 5 Tabellen. 1990. • 8.- (ISBN 3-506-72319-7)

Band 20: W. KREUER: Tagebuch der Heilig Land-Reise des Grafen Gaudenz von Kirchberg, Vogt von Matsch/Südtirol im Jahre 1470. - 349 Seiten, 3 beigelegte Karten, 7 Karten, 26 Abbildungen. 1990. • 23.- (ISBN 3-506-72320-0)

Band 21: J.-F. VENZKE: Beiträge zur Geoökologie der borealen Landschaftszone. Geländeklimatologische und pedologische Studien in Nord-Schweden. - 254 Seiten, 27 Fotos, 81 Abbildungen, 12 Tabellen. 1990. • 15.- (ISBN 3-506-72321-9)

Band 22: G. HENKEL (Hg.): Schadet die Wissenschaft dem Dorf? Vorträge und Ergebnisse des 7. Dorfsymposiums in Bleiwäsche vom 7.- 8. Mai 1990. - 150 Seiten. 1990. • 9.- (ISBN 3-506-72322-5)

Band 23: D. KELLETAT & L. ZIMMERMANN: Verbreitung und Formtypen rezenter und subrezenter organischer Gesteinsbildungen an den Küsten Kretas. - 163 Seiten, 37 Fotos, 45 Abbildungen, 7 Tabellen. 1991. • 14.- (ISBN 3-506-72323-5)

Band 24: G. HENKEL (Hg.): Der ländliche Raum in den neuen Bundesländern. Vorträge und Ergebnisse des 8. Essener Dorfsymposiums in Wilhelmsthal. Gemeinde Eckardtshausen in Thüringen (bei Eisenach) vom 25. - 26. Mai 1992. - 105 Seiten. 1992. • 11.- (ISBN 3-506-72324-3)

Band 25: J.-F. VENZKE (Hg.): Zur Ökologie und Gefährdung der borealen Landschaftszone. Beiträge zu einem borealgeographischen Kolloquium an der Universität Essen im Wintersemester 1993/94. - 173 S. 1994. • 21.- (ISBN 3-506-72325-1)

Band 26: G. HENKEL (Hg.): Außerlandwirtschaftliche Arbeitsplätze im ländlichen Raum. Vorträge und Ergebnisse des 9. Essener Dorfsymposiums in Bleiwäsche vom 9. - 10.5.1994. - 154 Seiten. 1995. • 35.- (ISBN 3-88474-291-4)

Band 27: H.-W. WEHLING: City im Wandel. Die Nutzungsstruktur der Essener Innenstadt 1995. -155 Seiten. 40 Abbildungen, 4 Faltkarten, 471 Tabellen. 1996. • 76.- (ISBN 3-88474-537-9) *vergriffen*

Band 28: G. HENKEL (Hg.): Das Dorf in Wissenschaft und Kunst. Vorträge und Ergebnisse des 10. Dorfsymposiums in Bleiwäsche vom 13. und 14. Mai 1996. - 106 Seiten, 35 Abbildungen. 1997. • 30.- (ISBN 3-88474-579-4) *vergriffen*

Band 29: G. SCHELLMANN: Jungkänozoische Landschaftsgeschichte Patagoniens (Argentinien). Andine Vorlandvergletscherungen, Talentwicklung und marine Terrassen. - 216 Seiten, 102 Abbildungen, 23 Tabellen, 36 Bilder, 7 Tabellenseiten im Anhang. 1998. • 81.- (ISBN 3-88474-671-5) *vergriffen*

Band 30: G. HENKEL (Hg.): 20 Jahre Dorferneuerung – Bilanzen und Perspektiven für die Zukunft. Vorträge des 11. Dorfsymposiums in Bleiwäsche vom 25. und 26. Mai 1998. - 128 Seiten, 21 Abbildungen, 8 Tabellen. 1999. • 20.- (ISBN 3-9803484-5-8)

Band 31: G. Henkel (Hg.): Das Dorf im Einflussbereich von Großstädten. - 134 Seiten, 43 Abbildungen, 15 Tabellen. 2000. • 20.- (ISBN 3-9803484-7-4)

Band 32: D. KELLETAT & G. SCHELLMANN (Hg.): Küstenforschung auf Zypern. Tsunamiereignisse und chronostratigraphische Untersuchungen. - 104 Seiten, 118 Abbildungen, 2 Tabellen. 2001. • 20.- (ISBN 3-9803484-8-2)

Band 33: A. SCHEFFERS: Paleotsunamis in the Caribbean. Field Evidences and Datings from Aruba, Curaçao and Bonaire. - 185 Seiten, 186 Abbildungen (davon 12 in Farbe), 18 Tabellen. 2002. • 34.- (ISBN 3-9803484-9-0)

Sonderband 1: Festausgabe aus Anlaß des 65. Geburtstages von Dieter WEIS. - 1986.

Sonderband 2: Essen im 19. und 20. Jahrhundert. Karten und Interpretationen zur Entwicklung einer Stadtlandschaft. Herausgegeben vom Vorstand der Geographischen Gesellschaft für das Ruhrgebiet, Essen. - 1990. *vergriffen*

Zu beziehen durch:

Band 1 - 25, ab Band 30 und Sonderbände:

Selbstverlag - Institut für Geographie
Universität Essen, Fachbereich 9, Universitätsstr. 15, 45117 Essen, e-mail: geographie@uni-essen.de, Fax: 0201/1832811

Band 26 - 29:

Klartext Verlag
Dickmannstr. 2, 45143 Essen, Tel.: 0201/ 86206- 31/32, Fax: 0201/86206- 22

ESSENER GEOGRAPHISCHE SCHRIFTEN

Band 1: F. SCHULTE-DERNE & H.-W. WEHLING: Atlas des Handwerks in Gelsenkirchen. - 128 Seiten, mit 10 Abbildungen, 53 Tabellen, 22 Farbfotos, 54 Farbkarten. Essen 1993. • 25.- (ISBN 3-9803484-0-7)

Band 2: W. KREUER: Imago Civitatis. Stadtbildsprache des Spätmittelalters. - 195 Seiten, mit 71 Abbildungen, 32 Faksimiles, Großformat im Schuber. Essen 1993. • 250.- / gefalzt: • 200.- (ISBN 3-9803484-1-5)

Band 3: W. KREUER.: Monumenta Cartographica. - 63 Seiten, 26 Abbildungen, 6 Kartentafeln, Großformat in Leinenkassette. Essen 1996. • 150.- (ISBN 3-9803484-3-1)

Zu beziehen durch:

Selbstverlag - Institut für Geographie
Universität Essen, Fachbereich 9, Universitätsstr. 15, 45117 Essen, e-mail: geographie@uni-essen.de, Fax: 0201/1832811

Sonstige Publikationen

D. KELLETAT (Ed.): Field Methods and Models to Quantify Rapid Coastal Changes. - Crete Field Symposium 1994, Program, Abstracts and Field Guide. - 80 Seiten, 27 Abbildungen und Karten, 3 Tabellen. Essen 1994. • 10.- (ISBN 3-9803484-2-3)

K. FEHN & H.W. WEHLING (Hg.): Bergbau und Industrielandschaften. - 327 Seiten, 79 Abbildungen und Karten, 8 Tabellen. Essen 1999. • 20.- (ISBN 3-9803484-6-6)

- ebenfalls über den Selbstverlag zu beziehen -